# Safety-Critical Systems eJournal

## Compilation Volume 1

## 2022

This collation page left blank intentionally.

Edited by John Spriggs
Cover image by Alex King

Published by the Safety-Critical Systems Club 2022

www.thescsc.org

SCSC-180

ISBN: 9798373163965

This collation page left blank intentionally.

# Contents

## About this Volume

This is the annual print compilation volume of all the issues of 2022 of the Journal of the Safety-Critical Systems Club (SCSC): ISSN 2753-6599 Volume 1.

The mission of the Safety-Critical Systems eJournal is to publish high-quality, peer-reviewed articles on the subject of systems safety. When we talk of systems, we mean not only the platforms, but also the people and their procedures that make up the whole. Systems Safety addresses those systems, their components, and the services they are used to provide. This is not a narrow view of system safety, our scope is wide and also includes safety-related topics such as resilience, security, public health and environmental impact.

Issue 1 and Issue 2 of Volume 1 were first issued on line as SCSC Publication Numbers SCSC-174 and SCSC-176; they are accessible as https://scsc.uk/scsc-174 and https://scsc.uk/scsc-176.

## Important Note

## About the Publisher

This journal is published by the Safety-Critical Systems Club (SCSC). The SCSC is a "Community Interest Company" (CIC), which is a special type of limited company that exists to benefit a community rather than private shareholders. The SCSC has:

- A 'community interest statement', explaining our plans;
- An 'asset lock' — a legal promise stating that our assets will only be used for our social objectives;
- A constitution; and
- Approval by the Regulator of Community Interest Companies.

Our community is that of Safety Practitioners. As a distinct legal entity the SCSC has more freedom and can legitimately do things such as make agreements with other bodies and own copyright on documents.

The SCSC began its work in 1991, supported by the UK Department of Trade and Industry and the Engineering and Physical Sciences Research Council. Since 1993 it has organised the annual Safety-Critical Systems Symposium (SSS) where leaders in different aspects of safety from different industries, including consultants, regulators and academics, meet to exchange information and experience, with the content published in this proceedings volume. The Club has been self-sufficient since 1994, and became a CIC in 2021.

The SCSC supports industry working groups. Currently there are active groups covering the areas of: Assurance Cases, Autonomous Systems, Data Safety, Multicore and Manycore, Ontology, Security Informed Safety, Service Assurance, Safety Culture and the Systems Approach to Safety of the Environment. These working groups provide a focus for discussions within industry and produce new guidance materials.

The SCSC maintains a website (thescsc.org, scsc.uk), which includes a diary of events, working group areas and club publications. It organises seminars, workshops and training on general safety matters or specific subjects of current concern. It produces a regular newsletter, Safety Systems, three times a year and now also a peer-reviewed journal. The journal is published on-line, as the Safety-Critical Systems eJournal, and comprises two issues a year, with an annual print volume: ISSN 2754-1118 (Online), ISSN 2753-6599 (Print).

**SCSC Mission:** To promote practical systems approaches to safety for technological solutions in the real world.

Where:

- "systems approaches" is the application of analysis tools, models and methods which consider the whole system and its components;
- "system" means the whole socio-technical system in which the solution operates, including organisational culture, structure and governance; and
- "technological solutions" includes products, systems and services and combinations thereof.

The Aims of the SCSC are:

1. To build and foster an active and inclusive community of safety stakeholders:
    a. "safety stakeholders" include practitioners (in safety specialisms and other disciplines involved in the whole lifecycle of safety related systems), managers, re searchers, and those involved in governance (including policy makers, law makers, regulators and auditors)

b. from across industry sectors, including new and non-traditional areas
c. recognising the importance of including and nurturing early career practitioners
d. working to remove barriers to inclusion in the community

2. To support sharing of systems approaches to safety:
    a. enabling wider application
    b. supporting continuing professional development
    c. encouraging interaction between early career and experienced practitioners
    d. using a variety of communication media and techniques to maximise coverage
    e. highlighting the lessons which can be learned from past experience
3. To produce consistent guidance for safety stakeholders where not already available.
    a. "consistent" meaning the guidance is consistent within itself, and with other guidance provided by SCSC; although SCSC will also aim to co-ordinate with external guidance this is more difficult to achieve
4. To influence relevant standards, guidance and other publications.
5. To work with relevant organisations to provide a co-ordinated approach to system safety.
6. To minimise our environmental impact wherever possible.

SCSC Membership may be either corporate or individual. Membership gives full web site access, the hardcopy newsletter, other mailings, and discounted entry to seminars, workshops and the annual Symposium. Corporate membership is for organisations that would like several employees to take advantage of the benefits of the SCSC. Different arrangements and packages are available. More information can be obtained at: scsc.uk/membership.

Also available is a short-term Publications Pass which, at very low cost, gives a month's access to all SCSC publications for non-members. Contact alex.king@scsc.uk for more details.

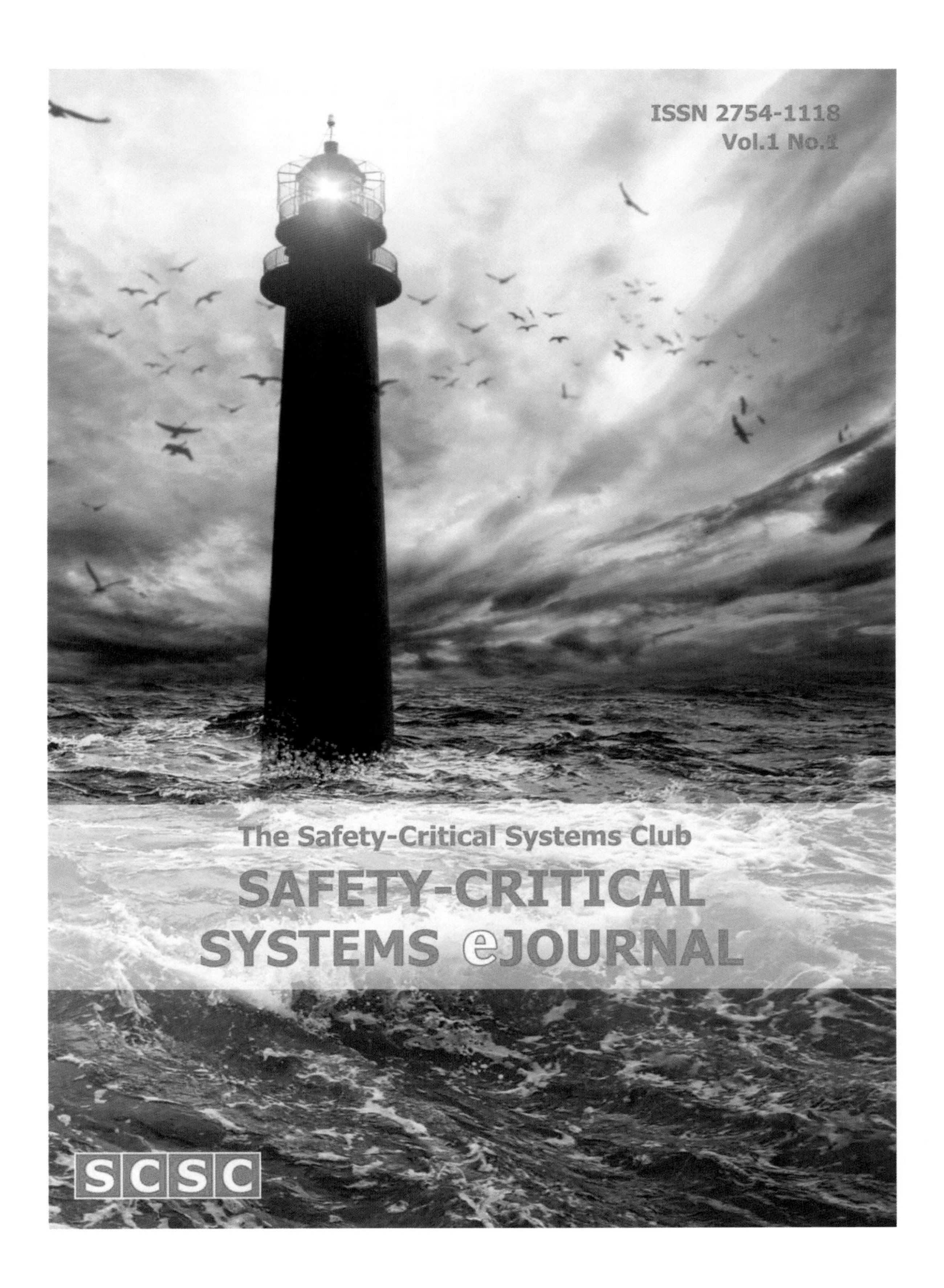
ISSN 2754-1118
Vol.1 No.1
The Safety-Critical Systems Club
SAFETY-CRITICAL
SYSTEMS eJOURNAL
SCSC

This collation page left blank intentionally.

# Welcome to the First Issue of a New Journal!

Welcome to the launch issue of the Safety-Critical Systems eJournal, which is published by the Safety Critical Systems Club, a UK Community Interest Company. Our mission is, "To publish high-quality, peer-reviewed articles on the subject of systems safety". By systems, we mean not only the platforms, but also the people and their procedures that make up the whole. Systems safety addresses those systems, their components, and the services they are used to provide. We do not have a narrow view of system safety, however; our scope is wide, and includes safety-related topics such as resilience, security, public health, and environmental impact.

This is not *purely* an academic journal presenting research results, we also intend to report upon practical aspects of systems safety; what works and what does not. The aim is to reflect how industrial practise is developing, for example how new standards are to be interpreted, or what new analysis techniques are being trialled. We are also interested in the past, how tools and techniques were justified for deployment, or significant lessons learned (and acted upon); and in the present, with topical or industry review articles, for example.

There will be two issues per volume, published in January and July of each year. In addition to the on-line presentation, each Volume of the journal will be made available in printed form both for libraries and for those of us who want a more-permanent record. The print volume will be published each December.

If you would like to submit a paper for a future issue, please see "Information For Authors" in the right-hand pane of the journal home page, https://scsc.uk/journal.

In this issue we have three papers:

- Rob Ashmore & James Sharp (UK) propose a set of generic assurance topics applicable to all types of programmable content for all types of platform, giving some novel examples;
- Dewi Daniels & Nick Tudor (UK) claim that many software reliability models do not provide results in which sufficient confidence can be placed, proposing an alternative approach; and
- Bruce Hunter (Australia) discusses how cyber-attacks, such as a ransomware attack on a business system, may also affect critical infrastructure, whether intentionally, or unintentionally; he considers cybersecurity threat mitigation, minimising the impact of attacks on safety systems.

My thanks go to the authors for contributing these papers, and also to the peer-reviewers (four per paper) for suggesting improvements. Apologies also to those reviewers who made some recommendations that were not taken up.

You may find some of this material controversial, or you may think that it does not go far enough. Subsequent issues of this journal will have provision for readers' letters to the Editor responding to individual papers.

For more resources addressing system safety, see the Club's web-site https://scsc.uk and please support us by becoming a member at https://scsc.uk/about?page=membership.

John Spriggs, Editor

January 2022

## About the Cover

The lighthouse, depicted here by Alex King as an island of stability in a stormy sea, represents a safety-critical system, which is intended to provide stable, trusted outcomes in a chaotic environment.

Lighthouses provide help to seafarers and have done so for thousands of years. Some warn of the presence of rocks or of other long-term hazards to navigation. Others are navigational aids and, with other cues, may be used to find a safe passage to harbour. Many perform both rôles; the lighthouse is thus itself (part of) a safety-critical system.

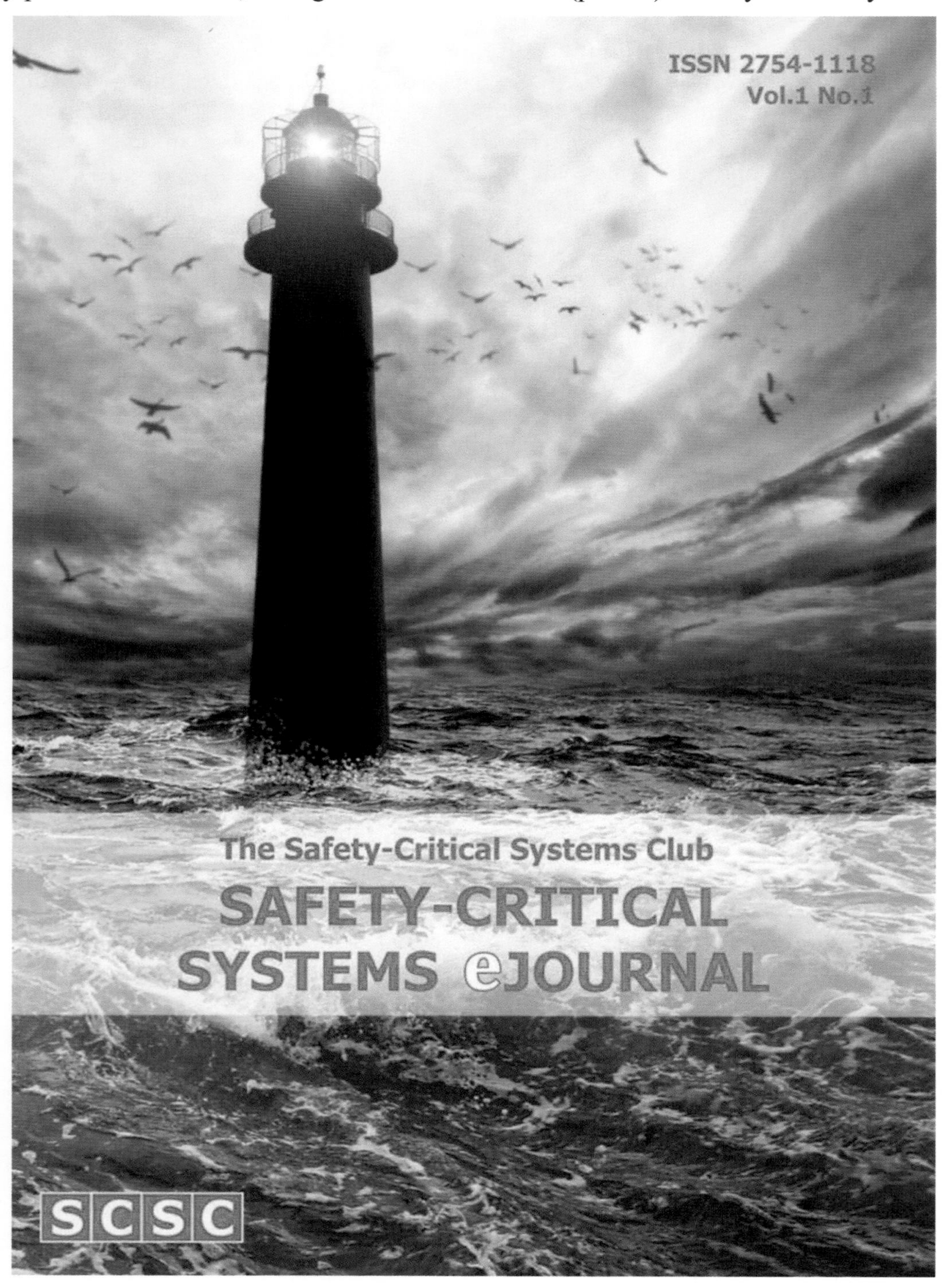

ISSN 2754-1118 (Online) — ISSN 2753-6599 (Print)

# Generic Assurance Topics for Any Type of Programmable Content

**Rob Ashmore and James Sharp**

Dstl, Portsdown West, UK

## Abstract

*An ever-growing range of technologies can be used to implement programmable content. Examples include highly complex System-On-Chip designs, which use traditional electronics, as well as approaches based on quantum technologies and biological systems. Standards and guidance are required to support the safe use of these, and other, emerging programmable technologies. However, technology-specific guidance is challenging to produce, especially in a timely manner. To help bridge this gap, we propose a set of generic assurance topics, which are applicable to all types of programmable content, introducing considerations based on the assurance of both a program, and the associated substrate. The topics are initially introduced by considering multi-core processors. Their application to alternate technologies is illustrated by considering electronic hardware tailored for machine learning, quantum computing and computation using a bio-based substrate.*

## 1 Introduction

### 1.1 Motivation

Safety-critical systems are making ever-increasing use of programmable content. This progress has been enabled by safety standards and guidance material. For example, RTCA/DO-178C (RTCA 2011a) addresses software, RTCA/DO-254 (RTCA 2000) addresses Complex Electronic Hardware (CEH) and both RTCA/DO-200 (RTCA 2015) and SCSC-127F (SCSC 2021) address data. Collectively, these examples cover three main aspects of programmability, specifically, software, hardware, and data. Whilst these examples are widely applicable, there is a growing recognition that technology-specific details are important. This is apparent, for example, in RTCA/DO-178C's supplements, e.g. RTCA/DO-331 (RTCA 2011b), which covers model-based development, and documents related to Multi-Core Processors (MCPs), e.g. CAST-32A (CAST 2016).

Programmable content technologies are developing rapidly. The associated need for specific, detailed information inevitably challenges the safety community. We could wait until the technology is well understood and key aspects have been codified as good practice, but this would mean standards significantly lag behind technology development. Another way of addressing the challenge would be to publish good practice rapidly, frequently updating it as new information becomes available. Neither of these approaches is desirable; standards should not lag technology, nor should they change too often (Johnson 2016).

Consequently, we frame a generic set of topics that an applicant would be expected to discuss with a regulator. The details of those discussions would be specific to the technologies involved in a particular application. Such discussions may be similar to those associated with Certification Review Items (CRIs), which are used in the aviation domain. Equivalently, the topics we identify may be an appropriate structure for CRIs related to programmable content.

Over time, as experience is gained, information may be collected and summarised into a more-traditional form of codified good practice. Although feasible, we recognise this approach has limitations. In particular, both the applicant and regulator need a sufficiently detailed understanding of the relevant technologies: this is not easy to achieve.

### 1.2 Related Work

The use of a generic set of topics is not a new concept. For example, there are already the "4+1 Software Safety Assurance Principles" (Hawkins et al. 2013). In addition, the Federal Aviation Administration (FAA) and the National Aeronautics and Space Administration (NASA) have described a set of overarching properties (Holloway 2019).

Our work differs from the software safety assurance principles in that it explicitly considers the substrate on which the software is running. In addition, it differs from the overarching properties in that it only considers programmable content, rather than a whole aircraft or an entire system. Consequently, our scope is broader than that of the software safety assurance principles and narrower than that of the overarching properties. Crucially, this choice of scope allows us to direct an appropriate amount of attention to novel types of programmable content.

## 2 Structure of Our Approach

### 2.1 Concepts

A program instantiates user-observable behaviour, where a "user" might be, for example, a human or another system component. Although the implementation may be complicated, we suggest that a program's intended behaviour is sufficiently constrained to allow it to be captured in a set of requirements that are meaningful to a user. This is often because a program has been developed to satisfy the needs of a particular user (or group of users). For clarity, we note that our concept of a program encapsulates both software and data.

A substrate provides the physical environment in which the program executes. The same substrate is expected to be able to support many different programs (not necessarily executing simultaneously). Equivalently, from the view of any individual program, the substrate is over-specified; a single program will only use a portion of the substrate's capabilities. In addition, both the physical nature of the substrate and the typically complex interaction between substrate and program result in substrate-level requirements being more detailed and more numerous than user-level program requirements. Consequently, we suggest that, typically, it is not possible for all aspects of a substrate's behaviour to be captured by requirements.

Without a substrate, a program is merely theoretical. Although some claims might be made about a program's behaviour, e.g. number of steps associated with a worst-case execution, these inevitably make assumptions about the substrate. Consequently, safety assurance of a program cannot be completed in isolation. Conversely, without a program, a substrate delivers no user-observable functionality, so safety assurance of a substrate in isolation is, at best, incomplete. Hence, safety assurance arguments related to programmable content are about a program (or a collection of programs) executing on a given substrate. This observation reflects current practice, e.g. RTCA/DO-178C's consideration of the target computer (RTCA 2011a).

To simplify presentation, we often refer to a single program, with a single developer. In almost all cases, there will be multiple programs and multiple developers. For example, in a traditional implementation, the substrate would be some form of processor, whereas the Operating System (OS) and applications would be programs.

The preceding discussion may suggest a clear and obvious separation between program and substrate. As the example of processor microcode illustrates, however, this is not always the case. Fundamentally, for us, the program contains things that, by design, would be expected to vary between programs and hence are under developer control. By that interpretation, microcode would be part of the substrate. Regardless of where the dividing line is drawn for any specific application of a particular technology, it is important that nothing is left unconsidered. This is another reason why, as noted earlier, assurance is about a program executing on a substrate.

## 2.2 Program Level Assurance

We are concerned with providing assurance that programmable content behaves as expected by the user, which may be a human or another system component. Our focus is on showing that the programmable content meets the expectations of the user, rather than demonstrating that the user's expectations will result in a safe system. Nevertheless, assurance of programmable content behaviour should be a useful claim, within a wider assurance argument, that a system is acceptably safe. As noted previously, a program's intended behaviour is captured in a set of requirements. Consequently, assurance needs to protect against all the following situations (Figure 1):

1. Where there is a difference in understanding of the requirements between the user and the program developer: colloquially, "**requirements are misunderstood**".
2. Where the program's behaviour is a strict subset of the requirements: "**some expected behaviour is not present**".
3. Where the program's behaviour is a strict superset of the requirements: "**some unexpected behaviour is present**".

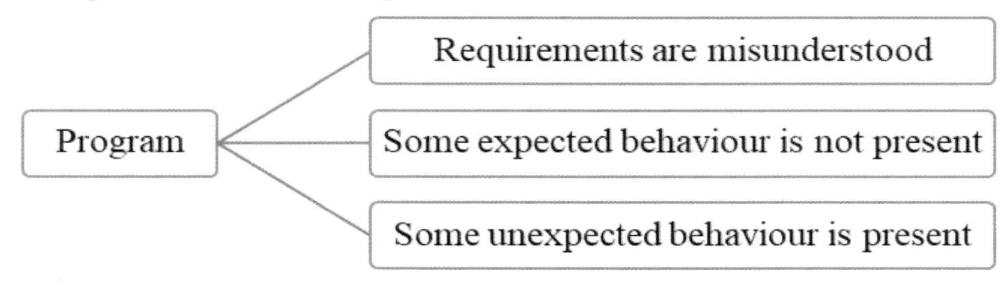


**Figure 1 ~ Program Level Situations to Protect Against**

It is helpful to reflect these behaviour-related outcomes in the more familiar context of traditional software.

In this case, protection against the first situation (**requirements are misunderstood**) is achieved by ensuring that requirements have particular characteristics, for example, accurate, complete, consistent, unambiguous and verifiable. Adopting a Requirements-Based Testing (RBT) approach, where the program and the tests are independently derived from the requirements, also helps protect against ambiguity in requirements.

RBT also protects against the second situation (**some expected behaviour is not present**). If the program's behaviour is verified against each requirement, then all expected behaviour should be present. This, of course, assumes that the verification methods are sound: control of the test environment and accuracy of tests cases are important considerations here.

Protection against the third situation (**some unexpected behaviour is present**) typically involves measuring code coverage using structural metrics, e.g. branch coverage, as part of an RBT endeavour. This approach is based on an implicit assumption that code structure (rather than, say, data) is the main way that different behaviours are instantiated. This approach also assumes that large-scale behaviour, as expressed in user-level requirements, can be decomposed to low-level behaviour, where coverage is typically measured. Robustness testing, e.g. checking for arithmetic overflow and exceeded frame times, also helps protect against some unexpected behaviour being present.

Formal methods (RTCA 2011c) provide an alternative approach to protecting against our three undesirable outcomes. In this approach, requirements are written in a formal language, guaranteeing an accurate, unambiguous, consistent (but not necessarily complete) description of intended behaviour. This description is refined, often through several stages, to formally demonstrate that the program instantiates the requirements. There have been several successful applications of formal methods; recent examples include a compiler (Leroy et al. 2016) and a microkernel (Klein et al. 2009).

Our three behaviour-related outcomes represent a slightly different way of considering program-level assurance. Although they cover similar themes, e.g. satisfaction of requirements, our outcomes provide an alternative perspective to the "4+1 Software Safety Assurance Principles" (Hawkins et al. 2013). Paradoxically, the widespread success of these principles, which summarise most current approaches to program-level safety, means they are not well-suited for our purpose. We believe they are, in many people's minds, wedded tightly to the development of traditional software, including the explicit, traceable, hierarchical decomposition of requirements. Consequently, they are not ideally suited for considering new types of programmable content (Ashmore and Lennon 2017). That said, our separation of program level and substrate level assurance means that the 4+1 Principles could be used for the former, with our topic areas being adopted for the latter. This might be an attractive approach when the program is largely traditional, with the novelty being in the substrate level.

In addition to the three behavioural outcomes noted above, there is also a need to consider protection against malicious intent. Different technologies may provide protection against certain types of malicious activity: the memory protection offered by the Rust programming language (Balasubramanian et al. 2017) is one example; the CHERI (Capability Hardware Enhanced RISC [Reduced Instruction Set Computer] Instructions) are another (Watson 2019). Most program developments are, however, subject to the same two vulnerabilities. Firstly, the insider threat, which may be partially protected against by review activities. Secondly, vulnerabilities inserted via the tool chain, e.g. (Thompson 2007), (Goodin 2017), and (Peisert et al. 2021), which may be partially (but only partially) protected against by obtaining professional-quality tools from reputable supply chains. Given the common nature of these vulnerabilities, for reasons of brevity, malicious activity at the program level is not discussed in detail in this paper. Nevertheless, it remains a very important issue.

### 2.3 Substrate Level Assurance

We adopt an expectation-based approach for substrate-level assurance. In particular, we are concerned with demonstrating that the substrate behaves as expected by the program developer.

As discussed previously, we contend that substrate-level behaviour cannot be meaningfully encapsulated (for a program developer) in a set of requirements. Inevitably, due to the *breadth* and *depth* that these requirements would have to cover, expectations only ever capture a subset.

The *breadth* issue arises because the substrate supports many different types of program. Consequently, an individual program will only use a fraction of the substrate's capability. A program developer will focus on those aspects of the substrate that appear relevant to their specific development. This understandable focusing inevitably produces a partial picture.

The *depth* issue arises because the substrate spans from a physical implementation to the level of abstraction used by the program developer: for traditional processors, the substrate spans from transistor-level properties of electrons to the Instruction Set Architecture (ISA). Across this scope, many aspects of physics can affect behaviour. For traditional electronics, single event upsets are an example that may arise naturally (Taber and Normand 1993), as are manufacturing and age-related issues (Dixit et al. 2021). There are also malicious attacks that exploit predictable outcomes of changes in the substrate's electronic environment, e.g. (Mutlu and Kim 2019) and (Murdock et al. 2020). Encapsulating a complete set of these effects in requirements seems, to us, an insurmountable challenge.

Typically, some form of on-target testing addresses common behaviour. Hence, the question of unexpected behaviour reduces to identification of potential edge cases. Ideally, these should be identified by a process that is rigorous, repeatable, and auditable. This is often achieved through the provision of some form of structure: the guidewords used in a hazard and operability study (HAZOP) are one example (Crawley and Tyler 2015). We propose the structure illustrated in Figure 2.

At the top level, this structure distinguishes between two classes. The first class covers cases where the program developer expects the substrate to behave in a manner that could be feasible but is not exhibited at the relevant time: colloquially, we term this a "**could be, but isn't**" (CBBI) behaviour. The second class covers cases where the developer expects a behaviour that "**could never be**" (CNB) delivered by the substrate. This distinction can be important because the former class (CBBI) can typically be controlled through configuration, whereas the latter class (CNB) cannot.

Note that the notion of expectation used here includes cases where the program developer expects something to happen, as well as cases where the program developer expects something not to happen. As such, it encapsulates situations where expected behaviour is absent, as well as situations where unexpected behaviour is present.

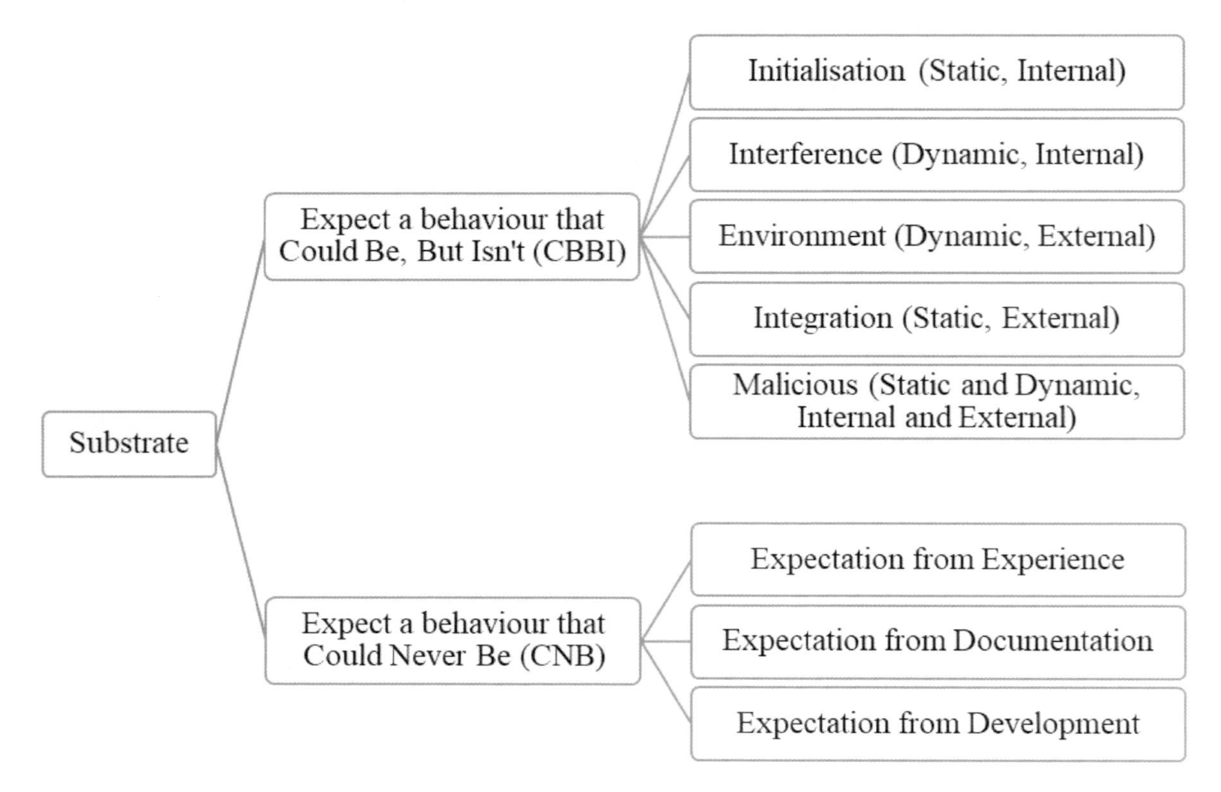


**Figure 2 ~ Substrate Level Situations to Protect Against**

Unexpected **CBBI** behaviours occur due to some influence on the substrate's behaviour. That influence may be **static,** i.e. remain fixed for the duration of program execution, or it may be **dynamic**. In addition, that influence may be **internal** to the substrate, or it may be **external**. The combination of these properties gives us four ways of identifying unexpected behaviour. These are illustrated in Table 1, with examples from the context of an MCP. These examples are provided solely to help illustrate our approach. As noted previously (in subsection 1.1), there are technology-specific standards for MCP, which are better suited to that specific technology than our more generic approach. Note that, as shown in Table 1, the "static, external" combination is addressed by (system) integration.

**Table 1 ~ Examples of CBBI Protections in an MCP Context**

| | **Static** | **Dynamic** |
|---|---|---|
| **Internal** | *Initialisation:* Ensuring the MCP configuration settings are as intended. | *Interference:* Mitigating the potential effects of shared resource use. |
| **External** | *Integration:* Introduction of a "safety net" (or "safety monitor") external to the MCP. | *Environment:* Protecting against single event upsets, for example, due to cosmic rays. |

In addition to the four combinations shown in Table 1, we explicitly consider a fifth item, specifically, **malicious** activity. Examples from the perspective of an MCP are shown in Table 2. This demonstrates that malicious activity can be associated with all four of the previously identified combinations, which may suggest that a specific malicious item is unnecessary. We have considerable sympathy with that view, especially as we strongly favour greater integration between, historically separate, safety and security activities. However, on balance, we believe the explicit security-related focus provided by the malicious item is beneficial. Note that we explicitly cover malicious activity for the substrate because it differs significantly between substrates. (At the level of detail considered in this paper, there is much more commonality in the types of malicious activity that are associated with the program level.)

**Table 2 ~ Examples of Malicious Activity for an MCP**

| | **Static** | **Dynamic** |
|---|---|---|
| **Internal** | *Malicious:* Hardware-based spyware that operates continuously. | *Malicious:* Hardware-based Trojan that is triggered by specific conditions. |
| **External** | *Malicious:* Undervolting a processor, to break secure enclave protection. | *Malicious:* Generation of external electro-magnetic fields to induce faults. |

For **CNB** behaviours, we seek to understand how the program developer may come to expect impossible ("**could never be**") substrate behaviour. We identify three potential sources for the mistaken expectation. Working from a more general perspective to a more specific one, these are:

- Expectations based on previous **experience**. For example, a developer used to working with single core processors may assume exclusive use (within a given time period) of peripherals. This might not be true for an MCP-based implementation.
- Expectations based on substrate **documentation**. These expectations could be mistaken, for example, because of errata or the presence of undocumented features (Domas 2017).
- Expectations based on **development** processes. Differences between the host, i.e. development, and target, i.e. operational, systems are a common source of mistaken expectations.

### 2.4 Outline Comparisons with Other Standards

We stress that our approach is not meant to replace existing, or emerging, technology-specific standards (or standards-like documents). Where technology-specific information is available, this should provide a more detailed description of relevant assurance topics (and how these may be addressed) than is possible for any generic approach. Technology-specific standards can describe both requirements and associated acceptable means of compliance. In contrast, our generic approach can only outline requirements, in the form of topic areas to be discussed.

As suggested above, our approach is intended to be used when no technology-specific standards are available. Since they satisfy different aims, a detailed comparison of our approach with existing technology-specific standards is not appropriate. Nevertheless, a top-level, outline comparison is considered beneficial, for two reasons. Firstly, it provides an additional description of our approach and how it may be used. Secondly, it provides confidence that the identified topic areas are both necessary and sufficient.

Consequently, our approach has been compared with CAST-32A (CAST 2016), which considers MCP, and the computation-level framework of SCSC-153A (SCSC 2020), which considers artificial intelligence implemented using machine learning techniques. Details of these outline comparisons are in Appendix A (Table 6 and Table 7, respectively).

The mapping to CAST-32A illustrates that all but one of CAST-32A's objectives can be mapped to our generic topic areas. The exception is that CAST-32A explicitly asks for an accomplishment summary, whereas communication mechanisms are not explicitly detailed in our approach. This mapping also shows that each of our areas maps to at least one CAST-32A objective.

The mapping to the SCSC-153A computation level shows that all of SCSC-153A's objectives can be mapped to our generic topic areas. It also shows that each topic area maps to at least one objective. It is noteworthy that our substrate topic areas, of which there are eight, correspond to only two of the nineteen SCSC-153A objectives. This reflects the greater prominence given to the substrate in our work, and is deemed appropriate, since this is an area of significant technological development.

The simple nature of these mappings means it would be inappropriate to read too much into their results. Nevertheless, they provide some confidence that the topic areas identified in our work are both necessary and sufficient for discussing assurance-related issues associated with new programmable content technologies. Further confidence is provided by the examples discussed in the following section.

# 3 Example Application

## 3.1 Example Programmable Content Technologies

To illustrate our topic areas, we apply them to three program/substrate combinations:

- An algorithm developed using Machine Learning (ML), running on a large-scale, complex System-On-Chip (SOC);
- An algorithm running on a quantum computer; and
- An algorithm running on a bio-based substrate.

The first example represents a technology that is available commercially now. The second and third technologies represent items that are being actively researched and where the associated substrates have very different properties to silicon-based electronics. We suggest that this range of examples (together with the previous MCP-related commentary) illustrates the generic nature of our approach. We note, however, that many other types of substrate could have been considered, e.g. neuromorphic processors (Davies et al. 2018), and memory-based compute (Bearden et al. 2020). The large number of potential examples is a key motivation for this paper.

The three examples are considered in the following subsections. Each subsection begins with a brief overview of the program and substrate. This is followed by a discussion of some of the key assurance points. Note that these discussions are not intended to be complete. Their purpose is not to act as a ready-made argument that can be deployed for any use of the associated substrate. Instead, the discussions are intended to demonstrate the way our generic assurance topics highlight key assurance aspects.

## 3.2 An Algorithm Developed using ML, Running on a SOC

**Overview:** This example uses Commercial Off-The-Shelf (COTS) hardware in the context of an autonomous vehicle. More specifically, the substrate is an NVIDIA Jetson AGX Xavier SOC (NVIDIA Corporation 2021). This includes: a 512-core Graphical Processing Unit (GPU); an 8-core Central Processing Unit (CPU); two Deep Learning

Accelerator (DLA) engines; a Very Long Instruction Word (VLIW) vision processor; Double Data Rate (DDR) 4 memory; and embedded Multi-Media Card (eMMC) storage.

We consider a program implemented as a Deep Neural Network (DNN), which performs object recognition on camera images. This program was developed using a combination of tools, including: the Jetson AGX Xavier Developer Kit; TensorRT; cuDNN; and TensorFlow. Training data was obtained from a fleet of camera-equipped vehicles. The program also includes the Board Support Package (BSP) and the OS, which in this case is Linux for Tegra (L4T).

Our considerations for this example draw heavily on (Ashmore and Sharp 2020), (Ashmore et al. 2021) and (SCSC 2020).

**Program Level:** For the purposes of this paper, it is assumed that the BSP and OS are largely traditional, so they are not considered further. In practice the BSP and OS are often an important part of the overall COTS package. As such, their influence on the behaviour of the substrate may require a thorough analysis.

ML is typically used to solve open problems, which do not have a complete set of accurate, consistent, and verifiable requirements. Indeed, if such a set of requirements is available then traditional software development techniques may be preferred (Salay and Czarnecki 2018).

When ML is used, protection against **requirements are misunderstood** may be achieved, at least in part, by:

- Ensuring training data is relevant to the object recognition task: for example, a training data set comprising German language road signs would not be relevant for an algorithm deployed on UK roads.
- Ensuring model behaviour is interpretable, including behaviour on a single input (local interpretability) and behaviour on classes of input (global interpretability).
- Ensuring independent verification activities provide suitable coverage of the inputs likely to be received during operational use.

Protection against **some expected behaviour is not present** may be achieved, at least in part, by:

- Ensuring measures of model performance adequately capture required behaviour. For example, some misclassifications may be more important than others: misclassifying a pedestrian as a road marking is likely to be worse than mistaking a 50 mph speed limit sign for a 30 mph limit. Consequently, a model that had similar misclassification rates, regardless of the classes involved, would not be demonstrating the expected behaviour.
- Ensuring test environments are sufficiently representative of the real world. This can be challenging, especially for open problems, where it is difficult to know which parts of reality need to be sufficiently represented.

Protection against **some unexpected behaviour is present** may be achieved, at least in part, by:

- Ensuring training data is suitably complete. For example, covering a sufficiently wide collection of classes, and sub-classes (Paterson and Calinescu 2019), of object types, observed in a sufficiently wide range of meteorological and illumination conditions.
- Ensuring testing on the target hardware covers, either exhaustively or via appropriate sampling, the training, test, and verification sets. Also, ensuring that this testing covers robustness issues, e.g. arithmetic overflow, numerical accuracy.
- Ensuring independent verification activities provide suitable coverage of inputs that may be received following failures elsewhere in the system, e.g. camera issues.

**Substrate Level:** The substrate is a COTS item. This means the programmable content developer is likely to, at best, have limited influence over its design features. It also means that the developer is likely to have only limited information about the SOC's implementation-level features. (A similar situation prevails for many modern processors, including MCPs, where detailed implementation-level feature information is unlikely to be available.)

From an **initialisation (CBBI - static, internal)** perspective, documenting and justifying the desired SOC configuration is important. This includes, but is not limited to, control of debug features and microcode updates.

From an **interference (CBBI - dynamic, internal)** perspective, runtime monitoring of the SOC's configuration is important (especially for software-configurable items). The potential effects of other software running alongside the program under test may also be important.

From an **environment (CBBI - dynamic, external)** perspective there may be a need to protect against the effect of active sensors on the autonomous vehicle, or other vehicles. For example, the combined effect of multiple, active sensors in a cluttered urban environment may need to be considered.

From an **integration (CBBI - static, external)** perspective, (system) integration activities providing for multiple processing channels, running on separate substrates, is an important consideration. For example, the object recognition channel (the subject of the current discussion) may be combined with a much simpler object detection channel. When multiple channels are used, the degree of independence needs to be carefully considered.

From a **malicious** perspective, the COTS nature of the substrate makes it extremely difficult to protect against the introduction of a hardware-based Trojan. Thinking about the standard cyber security triad of Confidentiality, Integrity and Availability (CIA), availability should be detectable and manageable by traditional safety measures, similar to a hardware failure of the SOC. Multiple processing channels should be used to detect loss of integrity. Confidentiality is very difficult to protect at the substrate-level. System-level architectural designs, which, from a confidentiality perspective, treat the SOC as an "untrusted box" may be an appropriate way of mitigating this risk.

From an **experience (CNB)** perspective, a developer may be unfamiliar with the combination of features available on the SOC. In particular, the combination of CPU, GPU and DLA may mean that the developer's program is executed on an unexpected part of the substrate. This risk may be exacerbated if different SOC workloads result in different execution patterns.

From a **documentation (CNB)** perspective, the possibility of undocumented features and document errata should be considered. The latter may be protected against by careful monitoring of information provided by the substrate manufacturer. Where specific features of the substrate are of particular importance, a "trust but verify" approach may be appropriate.

From a **development (CNB)** perspective, the SOC chosen for this example may be sufficiently powerful to support both development and operational use, thus removing host-target differences. Alternatively, these differences may be important, if development is conducted on a GPU, but operational inference is conducted on a DLA. One way of mitigating this concern would be to conduct all post-training evaluation on the intended target hardware.

### 3.3 An Algorithm Running on a Quantum Computer

**Overview:** We begin by noting that, if it is successfully commercialised (Dyakonov 2019), quantum computing is most likely to be applied as a co-processing technology, i.e. alongside traditional computers.

To illustrate the utility of quantum computing, it is helpful to consider different categories of problem (or application). For presentational reasons, we are deliberately imprecise: a more precise description is provided in (Gheorghiu et al. 2019). As shown in Table 3, we identify three separate classes, depending on whether a problem can easily be solved, and whether a solution can easily be checked, using traditional computers.

**Table 3 ~ Computation Classes Relevant to Quantum Computation**

| Using Traditional Computers ... | Class 1 | Class 2 | Class 3 |
|---|---|---|---|
| ... Easily Solved? | Yes | No | No |
| ... Easily Checked? | Yes | Yes | No |

For problems in Class 1, there is no need to invoke the additional complexity of a quantum computer. For problems in Class 2, provided the implications of a failed check can be handled safely, a significant amount of the assurance burden may be borne by using traditional computing to check a solution and, as such, assurance of the quantum portion may be less of a concern. Hence, problems in Class 1 and Class 2 are excluded from the current discussion.

To the best of our knowledge, the existence of Class 3 problems has not been proven. However, there are problems that credibly can be claimed as being in this class. One example requires calculation of a path that traverses a network (or graph) of a particular form (Childs et al. 2003).

A key aspect of a quantum computer is the notion of a "quantum bit", or qubit. Unlike a traditional bit, which holds a single value (either 0 or 1), a qubit holds a superposition of both 0 and 1; equivalently, a qubit holds a probability distribution over the space *{0, 1}*. Entanglement can be used to link qubits, so that the superposition of n qubits describes a probability distribution over a space containing 2n items. When a measurement is taken, the superposition collapses to a single item. The likelihood of receiving a particular measurement matches the associated probability encoded in the superposition.

Over recent years, there has been significant progress on quantum computing. For example, there is an open-source system, which includes a compiler, simulator and emulator (Steiger et al. 2018). This system includes a backend that links to cloud-based access to quantum computers, provided by IBM (IBM 2021). Other ways of interacting with these quantum computers are also available (IBM n.d.).

A simplified, but indicative, quantum computer system stack, based on (Fu et al. 2016), is illustrated in Table 4. This distinguishes program-level considerations and substrate-level considerations.

**Table 4 ~ Simplified Quantum Computer System Stack**

| Stack | Program / Substrate |
|---|---|
| Quantum Algorithm | Program |
| Programming Paradigm / Language | |
| Compiler | |
| Quantum Instruction Set Architecture | Substrate |
| Quantum Execution / Error Correction | |
| Quantum Chip | |

Given our current level of understanding, this example focuses on issues to be discussed, rather than on potential solutions to these issues.

**Program Level:** Our chosen example involves calculating a traversal path for graphs of a certain form. When considering **requirements are misunderstood**, it could be whether the program should calculate any path, or whether the shortest path should be calculated. As highlighted previously, quantum computing is inherently stochastic. Hence, requirements relating to the probability of obtaining the desired result (including whether results from sequential runs are statistically independent) may be misunderstood.

In terms of **some expected behaviour is not present**, test results may be influenced by inappropriate control of the test environment. This may be a particular concern if multiple tests are executed sequentially.

The program being considered is relatively simple in intent and, as such, offers little opportunity for cases where **some unexpected behaviour is present**. One possible example is different classes of network may be much easier to analyse. In some circumstances, significant changes in compute timing, for different classes of input, may be additional and unwanted behaviour.

**Substrate Level:** Quantum computation is an immature domain. Consequently, a specific example substrate is not discussed. Instead, we consider the quantum substrate in more general terms than that of the ML SOC.

From the perspective of **initialisation (CBBI - static, internal)**, we are concerned with the way that initial qubit values are established, as well as the way that qubit entanglements are created.

In terms of **interference (CBBI - dynamic, internal)**, decoherence and noise are significant challenges to the physical implementation of a quantum computer. These effects make it difficult to maintain specific values in, and entanglement between, qubits. Quantum Error Correction (QEC) methods are used to help mitigate these effects, at the expense of using a larger number of qubits.

With regards to **environment (CBBI - dynamic, external)**, current quantum computers require tight control of the environment, in particular to reduce the amount of environmental noise.

From the perspective of **integration (CBBI - static, external)**, the way that information is passed from a traditional computing resource to the quantum computer is of interest. Additionally, the number of times a quantum algorithm is repeated is another factor. Further, most QEC approaches rely on integration between a control mechanism implemented on a traditional computer and a quantum computer.

**Malicious** intervention into quantum cryptographic protocols is well-studied (Padmavathi et al. 2016). The potential vulnerabilities of quantum computers appear to be less well investigated. One obvious topic is an effective denial of service attack, caused by increasing the noise in the environment.

Quantum computer programs are based on a very different paradigm to traditional programs. For example, all steps in a quantum program need to be reversible, qubits cannot be copied, and no loops are permitted. (If necessary, loops can be achieved by repeated interaction between traditional and quantum computers, with the former handling the loop construct.) There are also cases where it is appropriate to use an undefined qubit in a calculation. These differences mean a developer may have unrealistic expectations based on previous **experience (CNB)**.

Inaccurate **documentation (CNB)** may create unrealistic expectations. One significant area involves understanding gate-level reliability, which depends on the QEC scheme employed, as well as the quantum computer's tolerance to external noise.

Unrealistic expectations may arise as part of **development (CNB)** due to inaccuracies in simulators and emulators used to debug quantum algorithms. Some of these, especially those relating to the final, pre-measurement superposition distribution, may be difficult to identify.

### 3.4 An Algorithm Running on a Bio-Based Substrate

**Overview:** This example considers molecular-based techniques, which exploit Chemical Reaction Networks (CRNs), implemented via Domain Strand Displacement (DSD) (Badelt et al. 2017). A brief introduction to CRNs is available in (Ashmore 2020).

A chemical reaction, in which reactants produce products at a given rate, is expressed in the general form:

$$Reactants \xrightarrow{Rate} Products \qquad \text{(Equation 1)}$$

Both *reactants* and *products* are more generally referred to as *species*. A CRN is a collection of related reactions.

In order to practically implement a CRN, fuel species are typically needed, e.g. to catalyse reactions, and waste species are often produced. Hence, a more complete description is:

$$Reactants + Fuel \xrightarrow{Rate} Products + Waste \qquad \text{(Equation 2)}$$

A number of implementations of CRNs have been demonstrated, including neural networks and solutions to the 3-SAT satisfiability problem (Winfree 2019).

For the purposes of our current discussion, we consider Equation 1 to be at the program level and Equation 2 to be at the substrate level; that level also includes, for example, the physical vessel in which the reactions occur.

Within this section, we assume the CRN is implemented in a bulk, well-mixed system rather than, for example, using a surface-based implementation (which, from our perspective, would be a different type of substrate). Where we need a specific example, we consider a combined CRN watchdog-oscillator system, where the watchdog can reset a failed oscillator (Ellis et al. 2019). However, much of the following focuses on generic CRN-related issues, rather than specific details of this example. This discussion is based on (Ellis et al. 2019), (Lutz et al. 2012) and (Johnson et al. 2019).

**Program Level:** In terms of **requirements are misunderstood**, CRNs are inherently stochastic, which needs to be appropriately reflected in the requirements. Requirements may also include implicit (and, consequently, misunderstood) assumptions regarding the fate, or future utility, of intermediate products. CRN behaviour is memoryless, stochastic, and asynchronous, so CRNs can be represented as Continuous Time Markov Chains (CTMCs). They are thus amenable to model checking, using tools like PRISM (Kwiatkowska et al. 2011), which can help detect ambiguous requirements.

Model checking can also provide a way of protecting against **some expected behaviour is not present**. CRNs also consume physical resources. In the case of our specific example, this may limit the number of times a broken oscillator may be restarted. Conversely, a user might expect unlimited restarts.

In CRNs, the concentration of a species, i.e. the amount present in the bulk mixture, can be viewed as analogous to a real-valued variable. More strictly, since concentrations cannot be negative, two species are needed to represent a real-valued variable: one for positive values and one for negative values. Including a set of reactions to cover negative values, when a variable can only take positive values, would be an example of **some unexpected behaviour is present** at the program level. Another example of unexpected behaviour could be including reactions to "tidy up" intermediary species, when not specified to do so in the requirements.

**Substrate Level:** In terms of **initialisation (CBBI - static, internal)**, key issues include the provision of sufficient quantities of reactant and fuel species. In addition, vessels used to contain the CRN implementation should be demonstrably free from contamination.

Concerning **interference (CBBI - dynamic, internal)**, in a bulk mixture all possible reactions occur simultaneously. This can make it challenging to schedule discrete steps within a program. Steps may be approximated with very different reaction rates (earlier steps having faster rates). Additionally, the products of a first step may be the reactants of a second step: this will ensure some of the first step completes before the second step begins; it will not ensure that the first step fully completes before the second step begins. Another approach involves using separate physical containers for each step: this obviously complicates the physical implementation (and integration). Another interference issue is unintended reactions, especially reactions that involve waste species.

From the perspective of the **environment (CBBI - dynamic, external)**, chemical reaction rates are sensitive to environmental conditions, like temperature and pressure. If these are not as intended, then the actual reaction rates may be very different to those anticipated. Additionally, the change in reaction rate may not be uniform across all reactions in a CRN, making sequencing of program steps even more challenging.

In terms of **integration (CBBI - static, external)**, a bio-based approach needs a mechanism for communicating the results of the computation. This is often achieved via molecules with different levels of luminescence. Using this type of approach to "read" the result of the computation is part of integration. For ongoing, continuous programs, e.g. monitors, integration may also involve ensuring there is a sufficient flow of fuel and reactants; removal of waste may also be a factor.

To the best of our knowledge, there has been little work on **malicious** interference with CRNs. Consequently, we merely highlight two theoretical possibilities. Firstly, we note that additional species could be inserted so that the program's behaviour changes after a certain number of oscillations (or, perhaps, a certain number of oscillator resets). Secondly, we note that the concentration of waste species could provide an adversary with valuable information, potentially undermining confidentiality.

From an **experience (CNB)** perspective, most programmers used to traditional electronics may be challenged by the notion that all possible reactions can occur simultaneously. The significant time taken for processing to complete (typically measured in hours) may be another relevant factor.

With regards to **documentation (CNB)**, the mapping from an abstract CRN, e.g. Equation 1, to an implementation CRN, e.g. Equation 2, means that, in essence, each substrate is program-specific. In traditional terms, the substrate is more like an Application-Specific Integrated Circuit (ASIC) than a general-purpose CPU. Hence, documentation relates to the way the relevant species are created and their resulting properties, e.g. amount of species fragments, or other unintended items.

In terms of **development (CNB)**, there are two main ways of analysing bulk mixtures. CTMCs account for stochastic behaviour and are amenable to model checking. Ordinary Differential Equations (ODEs) represent expected, average behaviour. This combination of methods may help protect against incorrect expectations. In addition, bisimulation approaches can provide confidence that an implementation CRN accurately reflects the intent of an abstract CRN (Johnson et al. 2019).

# 4 Summary

Table 5, overleaf, provides a very brief summary of the main points raised in the preceding examples. Depending on the nature of the example, these points may be mechanisms for providing assurance evidence, or issues that should be addressed by assurance evidence. For completeness, this table also includes a line on malicious activity at the program level.

The table demonstrates the potential applicability of our generic assurance topics to a wide range of programmable content. Whilst this is encouraging, more work is needed before these topics, broken down into program and application substrate, can become a formal recommendation. Completion of a number of specific worked examples, with sufficient detail to provide compelling assurance arguments, are an important part of that work.

Finally, motivated by the creation of specialised SOCs for ML tasks, and the broadening of computation away from traditional transistor-based architectures, we have highlighted the need for a set of generic assurance topics. These should enable a consistent approach to assuring elements within the ever-broadening domain of programmable content.

**Table 5 ~ Summary of Generic Assurance Topics and Chosen Examples**

| | | Developed Using ML, Running on a SOC | Algorithm Running on a Quantum Computer | Algorithm Running on a Bio-Based Substrate |
|---|---|---|---|---|
| Program | **Requirements are misunderstood** | Relevant training data; interpretable behaviour; independent verification | Requirements relating to probability of correct result; independence of sequential runs | Reflect stochastic nature; intermediate products; probabilistic model checking |
| | **Some expected behaviour is not present** | Adequate measures of model performance; sufficiently representative test environment | Control of the test environment, e.g. sequential tests | Probabilistic model checking; consumption of physical resources |
| | **Some unnecessary behaviour is present** | Sufficient testing (including robustness); independent verification | Significant, input-dependent changes in timing | Inclusion of species for negative values; tidy up of intermediate species |
| | **Malicious activity** | Insider threat; tool-based vulnerabilities | Insider threat; tool-based vulnerabilities | Insider threat; tool-based vulnerabilities |

| | | Developed Using ML, Running on a SOC | Algorithm Running on a Quantum Computer | Algorithm Running on a Bio-Based Substrate |
|---|---|---|---|---|
| Substrate | **Initialisation (CBBI - static, internal)** | Document and justify SOC configuration | Initial qubit values and entanglements | Sufficient quantities of species; vessels free from contamination |
| | **Interference (CBBI - dynamic, internal)** | Runtime monitoring; effects of co-hosted software | Decoherence; noise; QEC | Sequencing steps; unintended reactions |
| | **Environment (CBBI - dynamic, external)** | Active sensors on own, and other vehicles | Noise | Effects on reaction rates |
| | **Integration (CBBI - static, external)** | Multiple processing channels | Passing information from traditional to quantum; number of re-runs | Reading the results; input flow of species; removal of waste |
| | **Malicious activity** | Hardware-based Trojans; multiple processing channels; untrusted "closed box" architecture | Denial of service by increasing environment noise | Insertion of additional species; concentration of waste species |
| | **Experience (CNB)** | Combination of processing features | Reversible; no qubit copying; no loops | Simultaneous reactions; long processing time |
| | **Documentation (CNB)** | Undocumented features; document errata; trust but verify | Gate-level reliability; QEC scheme | Way species are created |
| | **Development (CNB)** | Host-target differences; train on target | Inaccuracies in simulators and emulators | CTMCs; ODEs; bisimulation |

## Disclaimers

This document is an overview of UK MOD sponsored research and is released for informational purposes only. The contents of this document should not be interpreted as representing the views of the UK MOD, nor should it be assumed that they reflect any current or future UK MOD policy. The information contained in this document cannot supersede any statutory or contractual requirements or liabilities and is offered without prejudice or commitment.

**Acknowledgments**

The authors gratefully acknowledge the comments provided by the anonymous reviewers. These led to a significantly improved paper.

**References**

Ashmore, R. (2020). Urban Maths - Best of Both Worlds. *Mathematics Today, June 2020.* Retrieved from: https://ima.org.uk/14267/urban-maths-the-best-of-both-worlds/. Accessed 14th December 2021.

Ashmore, R., & Lennon, E. (2017). Progress Towards the Assurance of Non-traditional Software. In Parsons, M., & Kelly, T. (Eds.) *Developments in System Safety Engineering, Proceedings of the 25th Safety-critical Systems Symposium, February 2017.* Independently Published.

Ashmore, R., & Sharp, J. (2020). Assurance Argument Elements for Off-the-shelf, Complex Computational Hardware. In Casimiro, A., Ortmeier, F., Bitsch, F., & Ferreira, P. (Eds.). *Computer Safety, Reliability, and Security: 39th International Conference, SAFECOMP 2020, Lisbon, Portugal, September 16–18, 2020, Proceedings* (pp. 260-269). Springer.

Ashmore, R., Calinescu, R., & Paterson, C. (2021). Assuring the Machine Learning lifecycle: Desiderata, methods, and challenges. *ACM Computing Surveys (CSUR), 54(5),* 1-39. https://doi.org/10.1145/3453444

Badelt, S., Shin, S. W., Johnson, R. F., Dong, Q., Thachuk, C., & Winfree, E. (2017). A General-Purpose CRN-to-DSD Compiler with Formal Verification, Optimization, and Simulation Capabilities. In Brijder, R., & Qian, L. (Eds). *DNA Computing and Molecular Programming: 23rd International Conference, DNA 23, Austin, TX, USA, September 24–28, 2017* (pp. 232-248). Springer, Cham, Switzerland. https://doi.org/10.1007/978-3-319-66799-7_15

Balasubramanian, A., Baranowski, M.S., Burtsev, A., Panda, A., Rakamarić, Z., & Ryzhyk, L. (2017). System Programming in Rust: Beyond Safety. In *Proceedings of the 16th Workshop on Hot Topics in Operating Systems, May 2017* (pp. 156-161). https://doi.org/10.1145/3102980.3103006

Bearden, S.R., Pei, Y.R., & Di Ventra, M. (2020) Efficient solution of Boolean satisfiability problems with digital memcomputing. *Scientific Reports, 10(1)*, 1-8. Retrieved from: https://www.nature.com/articles/s41598-020-76666-2 Accessed 20th January 2022.

CAST, the Certification Authorities Software Team (2016). *Multi-core Processors.* Position Paper CAST-32A, November 2016

Childs, A.M., Cleve, R., Deotto, E., Farhi, E., Gutmann, S., & Spielman, D. A. (2003). Exponential algorithmic speedup by a quantum walk. In *Proceedings of the thirty-fifth annual ACM symposium on Theory of computing, June 2003* (pp. 59-68) https://doi.org/10.1145/780542.780552

Crawley, F., & Tyler, B. (2015). *HAZOP: Guide to best practice*. Elsevier, 3rd edition (21 April 2015)

Davies, M., Srinivasa, N., Lin, T. H., Chinya, G., Cao, Y., Choday, S.H., Dimou, G., Joshi, P., Imam, N., Jain, S., & Liao, Y. (2018). Loihi: A Neuromorphic Manycore Processor with On-Chip Learning. *IEEE Micro, 38(1), January 2018,* 82-99 https://doi.org/10.1109/MM.2018.112130359

Dixit, H.D., Pendharkar, S., Beadon, M., Mason, C., Chakravarthy, T., Muthiah, B., & Sankar, S. (2021) *Silent Data Corruptions at Scale.* arXiv. Retrieved from: https://arxiv.org/pdf/2102.11245v1.pdf Accessed 20th January 2022.

Domas, C. (2017). *Breaking the x86 ISA.* Black Hat, USA. Retrieved from: https://www.blackhat.com/docs/us-17/thursday/us-17-Domas-Breaking-The-x86-Instruction-Set-wp.pdf Accessed 20th January 2022.

Dyakonov, M. (2019). When will useful quantum computers be constructed? Not in the foreseeable future, this physicist argues. Here's why: The case against: Quantum computing. *IEEE Spectrum, 56(3), March 2019,* (pp 24–29) https://doi.org/10.1109/MSPEC.2019.8651931

Ellis, S.J., Klinge, T.H., Lathrop, J.I., Lutz, J.H., Lutz, R.R., Miner, A.S., & Potter H.D. (2019). Runtime Fault Detection in Programmed Molecular Systems. *ACM Transactions on Software Engineering and Methodology (TOSEM), 28(2) April 2019, Article No.: 6,* (pp 1–20) https://doi.org/10.1145/3295740

Fu, X., Riesebos, L., Lao, L., Almudever, C.G., Sebastiano, F., Versluis, R., Charbon, E., Bertels, K. (2016). A heterogeneous quantum computer architecture. In *Proceedings of the ACM International Conference on Computing Frontiers, May 2016,* (pp 323–330) https://doi.org/10.1145/2903150.2906827

Gheorghiu, A., Kapourniotis, T., Kashefi, E. (2019). Verification of quantum computation: An overview of existing approaches. *Theory of Computing Systems, 63(4), May 2019* (pp.715-808) https://doi.org/10.1007/s00224-018-9872-3

Goodin, D. (2017). Apple scrambles after 40 malicious "XcodeGhost" apps haunt App Store. *Ars Technica.* Retrieved from: https://arstechnica.com/information-technology/2015/09/apple-scrambles-after-40-malicious-xcodeghost-apps-haunt-app-store/ Accessed 20th January2022.

Hawkins, R., Habli, I., & Kelly, T. (2013). The principles of software safety assurance. In *31st International System Safety Conference, 2013*. Retrieved from: https://www-users.cs.york.ac.uk/rhawkins/papers/HawkinsISSC13.pdf Accessed 20th January 2022.

Holloway, C. M. (2019). *Understanding the overarching properties.* NASA Technical Memorandum NASA/TM-2019-220292. Retrieved from: https://ntrs.nasa.gov/api/citations/20190029284/downloads/NASA-TM-2019-220292Replacement.pdf Accessed 20th January 2022.

IBM. (2021) *IBM Quantum Experience.* Retrieved from: http://research.ibm.com/quantum/, Accessed 14th December 2021.

IBM. (n.d.) *Qiskit: Open-Source Quantum Development*. Retrieved from: https://qiskit.org/. Accessed 14th December 2021.

Johnson, C. (2016). Role of Regulators in Safeguarding the Interface between Autonomous Systems and the General Public. In Hewett, J. (Ed.). *Proceedings of the 34th International System Safety Conference, Orlando, USA 8-12 August 2016* Retrieved from: http://www.dcs.gla.ac.uk/~johnson/papers/ISSC16/regulator.pdf Accessed 20th January 2022.

Johnson, R., Dong, Q., & Winfree, E. (2019). Verifying chemical reaction network implementations: a bisimulation approach. *Theoretical Computer Science, 765,* (pp.3-46). Retrieved from: https://authors.library.caltech.edu/96464/1/1-s2.0-S0304397518300136-main.pdf Accessed 20th January 2022.

Klein, G., Elphinstone, K., Heiser, G., Andronick, J., Cock, D., Derrin, P., Elkaduwe, D., Engelhardt, K., Kolanski, R., Norrish, M., & Sewell, T. (2009*)*. seL4: Formal verification of an OS kernel. In *Proceedings of the ACM SIGOPS 22nd Symposium on Operating Systems Principles, October 2009,* (pp. 207-220) https://doi.org/10.1145/1629575.1629596

Kwiatkowska, M., Norman, G., & Parker, D. (2011). PRISM 4.0: Verification of probabilistic real-time systems. In Gopalakrishnan, G., & Qadeer, S. *Computer Aided Verification: 23rd International Conference, CAV 2011, Snowbird, UT, USA, July 14-20, 2011. Proceedings*. (pp. 585-591). Springer. Berlin, Heidelberg, Germany.

Leroy, X., Blazy, S., Kästner, D., Schommer, B., Pister, M., & Ferdinand, C. (2016). CompCert-a formally verified optimizing compiler. In *Proceeding of the 8th European Congress on Embedded Real Time Software and Systems, ERTS 2016, Toulouse, France* Retrieved from: https://www.researchgate.net/publication/293814383_CompCert_-_A_Formally_Verified_Optimizing_Compiler Accessed 20th January 2022.

Lutz, R. R., Lutz, J. H., Lathrop, J. I., Klinge, T. H., Mathur, D., Stull, D. M., Bergquist, T. G., & Henderson, E. R. (2012). Requirements analysis for a product family of DNA nanodevices. In *2012 20th IEEE International Requirements Engineering Conference (RE)* (pp. 211-220). https://doi.ieeecomputersociety.org/10.1109/RE.2012.6345806

Murdock, K., Oswald, D., Garcia, F.D., Van Bulck, J., Gruss, D., & Piessens, F. (2020). Plundervolt: Software-based fault injection attacks against Intel SGX. In *2020 IEEE Symposium on Security and Privacy (SP)* (pp. 1466-1482). https://doi.ieeecomputersociety.org/10.1109/SP40000.2020.00057

Mutlu, O., & Kim, J. S. (2019). RowHammer: A retrospective. *IEEE Transactions on Computer-Aided Design of Integrated Circuits and Systems, 39(8), August 2020.* 1555-1571 https://doi.org/10.1109/TCAD.2019.2915318

NVIDIA Corporation. (2021). *Jetson AGX Xavier Developer Kit.* Retrieved from: https://developer.nvidia.com/embedded/jetson-agx-xavier-developer-kit. Accessed 14th December 2021.

Padmavathi, V., Vardhan, B. V., & Krishna, A. V. N. (2016). Quantum cryptography and quantum key distribution protocols: a survey. In *2016 IEEE 6th International Conference on Advanced Computing (IACC)* (pp. 556-562). https://doi.org/10.1109/IACC.2016.109

Paterson, C., & Calinescu, R. (2019). *Detection and mitigation of rare subclasses in neural network classifiers.* arXiv preprint arXiv:1911.12780, 2019. Retrieved from: https://www.researchgate.net/profile/Colin-Paterson-4/publication/337671290_Detection_and_Mitigation_of_Rare_Subclasses_in_Neural_Network_Classifiers/links/5df770984585159aa4809742/Detection-and-Mitigation-of-Rare-Subclasses-in-Neural-Network-Classifiers.pdf Accessed 20th January 2022.

Peisert, S., Schneier, B., Okhravi, H., Massacci, F., Benzel, T., Landwehr, C., Mannan, M., Mirkovic, J., Prakash, A., & Michael, J. B. (2021). Perspectives on the SolarWinds Incident. *IEEE Security & Privacy, 19(02)*, (pp. 7-13). Retrieved from: https://www.computer.org/csdl/magazine/sp/2021/02/09382367/1saZVPHhZew Accessed 20th January 2022.

RTCA. (2000). *Design Assurance Guidance for Electronic Hardware*. RTCA/DO-254, RTCA, Inc. Also available as EUROCAE Document ED-80.

RTCA. (2011a). *Software Considerations in Airborne Systems and Equipment Certification*. RTCA/DO-178C, RTCA, Inc. Also available as EUROCAE Document ED-12C.

RTCA. (2011b). *Model-Based Development Supplement*. RTCA/DO-331, RTCA, Inc. Also available as EUROCAE Document ED-218.

RTCA. (2011c). *Formal Methods Supplement to DO-178C and DO-278A*. RTCA/DO-333, RTCA, Inc. Also available as EUROCAE Document ED-216.

RTCA. (2015). *Standards for Processing Aeronautical Data*. RTCA/DO-200, RTCA, Inc. Also available as EUROCAE Document ED-76.

Salay, R., & Czarnecki, K. (2018). *Using machine learning safely in automotive software: An assessment and adaption of software process requirements in ISO 26262*. arXiv preprint arXiv:1808.01614, 2018 Retrieved from: https://arxiv.org/ftp/arxiv/papers/1808/1808.01614.pdf Accessed 20th January 2022.

SCSC, Safety Critical Systems Club C.I.C. (2020). *Safety assurance objectives for autonomous systems*, SCSC-153A. Retrieved from: https://scsc.uk/r153A:1 Accessed 20th January 2022.

SCSC, Safety Critical Systems Club C.I.C. (2021). *Data Safety Guidance*. SCSC-127F. Retrieved from: https://scsc.uk/r127F:1 Accessed 20th January 2022.

Steiger, D. S., Häner, T., & Troyer, M. (2018). ProjectQ: an open source software framework for quantum computing. *Quantum, 2,* p.49 https://doi.org/10.22331/q-2018-01-31-49

Taber, A., & Normand, E. (1993). Single event upset in avionics. *IEEE Transactions on Nuclear Science, vol. 40, no. 2, pp. 120-126, April 1993,* https://doi.org/10.1109/23.212327

Thompson, K. (2007). Reflections on trusting trust. In *ACM Turing Award Lectures* (p. 1983), January 2007. Association for Computing Machinery. New York, USA https://doi.org/10.1145/1283920.1283940

Watson, R. N. M., Moore, S. W., Sewell, P., & Neumann, P. G. (2019). *An introduction to CHERI*. University of Cambridge, Computer Laboratory, Technical Report Number 941, UCAM-CL-TR-941, ISSN 1476-2986

Winfree, E. (2019). Chemical reaction networks and stochastic local search. In *International Conference on DNA Computing and Molecular Programming* (pp. 1-20). Springer. Cham Switzerland.

# Appendix A. Outline Comparison Mappings

Table 6, below, provides an outline mapping between the objectives listed in CAST-32A and the generic topic areas identified in this paper. Table 7 provides a similar mapping between the objectives in the computation-level framework of SCSC-153A and the topic areas of this paper.

**Table 6 ~ Outline Mapping from CAST-32A Objectives to Generic Topic Areas**

| CAST-32A Objective | Program | | | Substrate | | | | | | | | Count |
|---|---|---|---|---|---|---|---|---|---|---|---|---|
| | Requirements are misunderstood | Some expected behaviour is not present | Some unexpected behaviour is present | CBBI Initialisation (Static, Internal) | CBBI Interference (Dynamic, Internal) | CBBI Environment (Dynamic, External) | CBBI Integration (Static, External) | CBBI Malicious | CNB Expectation from Experience | CNB Expectation from Documentation | CNB Expectation from Development | |
| MCP_Planning_1 | | | | X | X | | X | | | X | X | **5** |
| MCP_Resource_Usage_1 | | | | X | X | | | | | | | **2** |
| MCP_Resource_Usage_2 | | | | | X | X | | X | | | | **3** |
| MCP_Planning_2 | | | | X | X | | X | | | X | | **4** |
| MCP_Resource_Usage_3 | | | | | X | | | | | X | | **2** |
| MCP_Resource_Usage_4 | | | | | X | | X | | | | | **2** |
| MCP_Software_1 | X | X | X | | X | | | | | | | **4** |
| MCP_Software_2 | X | X | X | | X | | | | X | X | | **6** |
| MCP_Error_Handling_1 | | | | | | | X | X | | | | **2** |
| MCP_Accomplishment_Summary_1 | | | | | | | | | | | | **0** |
| **Count** | **2** | **2** | **2** | **3** | **8** | **1** | **4** | **2** | **1** | **4** | **1** | **-** |

**Table 7 ~ Outline Mapping from SCSC-153A Computation-Level Objectives to Generic Topic Areas**

| SCSC-153A Computation-Level Objective | Program | | | Substrate | | | | | | | | Count |
|---|---|---|---|---|---|---|---|---|---|---|---|---|
| | Requirements are misunderstood | Some expected behaviour is not present | Some unexpected behaviour is present | CBBI Initialisation (Static, Internal) | CBBI Interference (Dynamic, Internal) | CBBI Environment (Dynamic, External) | CBBI Integration (Static, External) | CBBI Malicious | CNB Expectation from Experience | CNB Expectation from Documentation | CNB Expectation from Development | |
| COM 1-1: Data is acquired and controlled appropriately. | X | | | | | | | | | | | **1** |
| COM1-2: Pre-processing methods do not introduce errors. | | X | X | | | | | | | | | **2** |
| COM1-3: Data captures the required algorithm behaviour. | | X | | | | | | | | | | **1** |
| COM1-4: Adverse effects arising from distribution shift are protected against. | | | X | | | | | | | | | **1** |
| COM2-1: Functional requirements imposed on the algorithm are defined and satisfied. | X | X | X | | | | | | | | | **3** |
| COM2-2: Non-functional requirements imposed on the algorithm are defined and satisfied. | X | X | X | | | | | | | | | **3** |
| COM2-3: Algorithm performance is measured objectively. | | X | | | | | | | | | | **1** |
| COM2-4: Performance boundaries are established and complied with. | | X | | | | | | | | | | **1** |
| COM2-5: The algorithm is verified with an appropriate level of coverage. | | X | X | | | | | | | | | **2** |
| COM2-6: The test environment is appropriate. | | X | | | | | | | | | | **1** |
| COM2-7: Each algorithm variant is tested appropriately. | | X | X | | | | | | | | | **2** |

| **SCSC-153A Computation-Level Objective** | **Program** | | | **Substrate** | | | | | | | | **Count** |
|---|---|---|---|---|---|---|---|---|---|---|---|---|
| | Requirements are misunderstood | Some expected behaviour is not present | Some unexpected behaviour is present | CBBI Initialisation (Static, Internal) | CBBI Interference (Dynamic, Internal) | CBBI Environment (Dynamic, External) | CBBI Integration (Static, External) | CBBI Malicious | CNB Expectation from Experience | CNB Expectation from Documentation | CNB Expectation from Development | |
| COM3-1: An appropriate algorithm type is used. | | X | X | | | | | | | | | **2** |
| COM3-2: Typical errors are identified and protected against. | | | X | | | | | | | | | **1** |
| COM3-3: The algorithm's behaviour is explainable. | X | X | X | | | | | | | | | **3** |
| COM3-4: Post-incident analysis is supported. | | X | X | | | | | | | | | **2** |
| COM4-1: The software is developed and maintained using appropriate standards. | X | X | X | | | | | | | | | **3** |
| COM4-2: Software misbehaviour does not result in incorrect outputs from the algorithm. | | | X | | | | | | | | | **1** |
| COM5-1: Appropriate computational hardware standards are employed. | | | | X | X | X | X | X | X | X | X | **8** |
| COM5-2: Hardware misbehaviour does not result in incorrect outputs from the algorithm. | | | | | X | X | | X | | | | **3** |
| **Count** | **5** | **13** | **12** | **1** | **2** | **2** | **1** | **2** | **1** | **1** | **1** | **-** |

ISSN 2754-1118 (Online) — ISSN 2753-6599 (Print)

# Software Reliability and the Misuse of Statistics

**Dewi Daniels[1], and Nicholas (Nick) Tudor[2]**

1. Software Safety Limited, Trowbridge, UK
2. D-RisQ Ltd, Malvern, UK

## Abstract

*Many papers have been written on software reliability. The claim is made that failures of software-based systems occur randomly and that statistical techniques used to predict random hardware failure rates can also be used to predict software failure rates. This claim has not been challenged in any academic papers, though it is treated with suspicion by many practising engineers. As a result, the applicability of these statistical techniques has been accepted in some standards, such as IEC 61508, but has been rejected in others, such as RTCA/DO-178C. It is more important than ever to understand whether this claim is true. There is strong lobbying from industry to allow software not developed to any standard to be used for safety critical applications provided it has sufficient product service history. The European Union Aviation Safety Agency (EASA) is promoting dissimilar software in the belief that using two or more independent software teams will deliver ultra-high levels of software reliability. Software defects are different from random hardware failures and need to be treated differently. This paper argues that the techniques used for statistical evaluation of software make unwarranted assumptions about software and lead to overly optimistic predictions of "software failure rates". This paper concludes that many software reliability models do not provide results in which confidence can be placed. Instead, this paper proposes an alternative way forward that does provide evidence that software is safe for its intended use before it enters service.*

## 1 Introduction

Many papers have been written on software reliability, including (Littlewood and Verrall 1973) and (Ladkin and Littlewood 2016). The claim is made that the same statistical techniques used to predict random hardware failure rates can also be used to predict software failure rates. While (Ladkin and Littlewood 2016) admits that "*It is true that software fails systematically, in that, if a program fails in certain circumstances, it will always fail when those circumstances are exactly repeated*", the authors go on to claim that, "*there is uncertainty about when a program will receive an input that will cause it to fail*". They therefore conclude that software failures form a stochastic (random) process and that probabilistic models can be applied to software failures. A view is that as software is a component of a system and all components contribute to the reliability of the system, that software therefore has to have a failure rate. This reflects the demands of some reliability engineers using techniques such as Failure Mode and Effects Criticality Analysis (FMECA) and therefore also demands to have failure mechanisms for software.

The applicability of these statistical techniques is accepted by some industry standards, such as IEC 61508-3 (IEC 2010a), but rejected by others, such as RTCA/DO-178C (RTCA 2011).

There is pressure from industry to allow commercial-off-the-shelf (COTS) and open source software to be accepted for use in safety critical systems based on product service history. For example, RTCA SC-240/EUROCAE WG-117 has been directed by RTCA Program Management Committee (PMC) and the EUROCAE Council (EC) to develop guidance on the integration of COTS, open source, and service history into software for airborne systems (RTCA 2021).

Furthermore, the European Union Aviation Safety Agency (EASA) is advocating dissimilar software development in the belief that the use of multiple, independent software teams will prevent common mode software failures (EASA 2014).

# 2 Hardware Reliability Modelling

The mechanical domain successfully uses statistical models. However, that world is markedly different from the one occupied by software. There are no similar failure mechanisms for software: there is no corrosion, fatigue, or fungus, let alone vagaries in manufacturing, which engineers account for by building in safety factors. It has been shown over many years that the use of modelling and statistics in techniques such as FMECA give well understood and predictable outcomes. These are models built up from knowledge of individual component failure mechanisms, their likelihood, and contribution to the overall system reliability. Looking for the contribution from the software domain, some mechanical engineers wish to have a number for reliability of the software component. Unfortunately, a response that it is either a '1' or a '0', does not tend to satisfy, so seeking a statistical approach to determining the contribution of software to overall system reliability is understandable perhaps, if one thinks of software as a component of a system, like all other components; unfortunately, it is not like other components, and needs to be treated differently, as we shall now discuss.

It should be noted that complex electronic hardware (CEH), on the other hand, can fail in very similar ways to software. Indeed, writing in VHDL (VHSIC [Very High-Speed Integrated Circuits] Hardware Description Language) is very similar in many ways to software programming. The issues raised in this paper could therefore be applied equally to CEH.

# 3 Software Reliability Modelling

## 3.1 Software Reliability in the Standards

### *3.1.1 Software Reliability in RTCA/DO-178C*

RTCA/DO-178C (RTCA 2011) warns, "*Development of software to a software level does not imply the assignment of a failure rate for that software. Thus, software reliability rates based on software levels cannot be used by the system safety assessment process in the same way as hardware failure rates*". Concerning software reliability models, it states, "*Many methods for predicting software reliability based on developmental metrics have been published, for example, software structure, defect detection rate, etc. This document does not provide guidance for those types of methods, because at the time of writing, currently available methods did not provide results in which confidence can be placed*".

### *3.1.2 Software Reliability in IEC 61508*

IEC 61508-3 (IEC 2010a) recommends probabilistic testing (highly recommended for software aspects of system safety validation at SIL 4). IEC 61508-7 (IEC 2010b) recognises that it is very difficult to demonstrate ultra-high levels of reliability using these techniques. (Littlewood and Strigini 1993) acknowledge that, where ultra-high dependability is required for software-based systems, it is not possible to confirm that a sufficiently high, numerically expressed dependability has been achieved.

Annex D of (IEC 2010b) describes a probabilistic approach to determining software integrity for pre-developed software. It is noted that Annex D is informative rather than normative. Table below shows the number of failure-free demands experienced or hours of failure-free operation needed to qualify for a particular safety integrity level according to Annex D.

**Table 1 ~ Necessary History from IEC 61508-7 Annex D**

| **SIL** | **Low demand mode of operation** | **Number of treated demands** | | **High demand or continuous mode of operation** | **Hours of operation in total** | |
|---|---|---|---|---|---|---|
| | (Probability of failure to perform its design function on demand) | $1-\alpha = 0.99$ | $1-\alpha = 0.95$ | (Probability of a dangerous failure per hour) | $1-\alpha = 0.99$ | $1-\alpha = 0.95$ |
| 4 | $\geq 10^{-5}$ to $< 10^{-4}$ | $4.6 \times 10^{5}$ | $3 \times 10^{5}$ | $\geq 10^{-9}$ to $< 10^{-8}$ | $4.6 \times 10^{9}$ | $3 \times 10^{9}$ |
| 3 | $\geq 10^{-4}$ to $< 10^{-3}$ | $4.6 \times 10^{4}$ | $3 \times 10^{4}$ | $\geq 10^{-8}$ to $< 10^{-7}$ | $4.6 \times 10^{8}$ | $3 \times 10^{8}$ |
| 2 | $\geq 10^{-3}$ to $< 10^{-2}$ | $4.6 \times 10^{3}$ | $3 \times 10^{3}$ | $\geq 10^{-7}$ to $< 10^{-6}$ | $4.6 \times 10^{7}$ | $3 \times 10^{7}$ |
| 1 | $\geq 10^{-2}$ to $< 10^{-1}$ | $4.6 \times 10^{2}$ | $3 \times 10^{2}$ | $\geq 10^{-6}$ to $< 10^{-5}$ | $4.6 \times 10^{5}$ | $3 \times 10^{6}$ |

**NOTE 1** $1-\alpha$ represents the confidence level.

**NOTE 2** See [IEC 61508-7] D.2.1 and D.2.3 for prerequisites and details of how this table is derived

## 3.2 Applicability of Models

While models can be useful in various circumstances, users have to be careful to ensure that the chosen model is applicable and that the limitations of the model are understood when drawing conclusions. (Mandelbrot and Hudson 2004) predicted the stock market crash that occurred in 2008. (Mandelbrot and Hudson 2004) claimed that the mathematical models used were flawed and that it was mistaken to assume that the normal distribution was a useful model for tracking price changes in the stock markets. Most economists responded that independence and normality are just assumptions that help simplify the mathematics. However, the inappropriate application of the normal distribution underestimated the probability that many borrowers would default on their subprime mortgages at the same time.

One must be careful of the "tail" in the models. If the application of the model underestimates the probability of an extreme event, then the consequences of decisions made to defend against such an event could be far reaching. (Mandelbrot and Hudson 2004) claimed that stock market prices follow a power law rather than the normal distribution. Power laws have fatter tails than the normal distribution. A statistical model based on the normal distribution would have underestimated the probability of unlikely events such as stock market crashes.

There are many examples of incorrect application of modelling to problems. Indeed, a quote generally attributed to George Box is, "All models are wrong, but some are useful", which tries to explain that models are not the real artefact but may nevertheless be useful. The scientist, engineer, forecaster must therefore ensure that the assumptions behind the use of the model are the right assumptions to be making and which must be understood before valid conclusions can be made. Without this understanding, a false premise can lead to almost any conclusion one would like, or indeed, not like.

## 3.3 Assumptions

### *3.3.1 Preamble*

Many software reliability models assume that software execution is a Bernoulli process. It is therefore assumed that:

1. Executing a software program results in one of two outcomes, which we term Success or Failure; and
2. The probability of Success is the same every time the software is executed.

The first of these assumptions seems a reasonable assumption. The software either produces the correct output or it does not. We accept it is not always easy to determine whether an output is correct or not, but we will assume for now that we can do this. The second of these assumptions is an unwarranted assumption in most cases, as we explore in the following examples.

### *3.3.2 Example 1: State*

Most software programs contain state. This could be in the form of a variable stored in Random Access Memory (RAM), a record stored on a Solid-State Drive (SSD), a Hard Disk Drive (HDD), in the Cloud, or even as the contents of a memory cache. State essentially introduces bias into the system. A fair coin or die does not have "memory". Software state is worse than a biased coin because the bias could change from one trial to the next. Suppose a software program increments an unsigned 16-bit counter. When the counter reaches 65,535, it wraps around to zero, causing the program to produce the wrong output. The probability of success is not the same every time the software is executed. Assuming there are no other defects, the probability of success the first 65,535 times the software program is executed is one; the probability of success the 65,536$^{th}$ time the software program is executed is zero. If we had executed the software program 65,535 times and observed it produce the correct output each time, we would have a high degree of confidence that the software program is correct, yet it will fail the very next time it is invoked.

The presence of state means that whether invoking a software program results in success or failure depends not just on the inputs presented at that time, but on previous invocations of the software. If you are worrying about cache contents, then it also depends on other software that is running on the same hardware (even if we're using time slicing!). If the software halts the processor or enters an infinite loop, then all subsequent invocations will fail until the software is reset. More subtly, a previous invocation of the software could have resulted in the state changing in such a way that a future invocation of the software will fail. This means that the probability of success is not the same every time the software is executed.

### *3.3.3 Example 2: Environment*

The mathematical models assume that the probability of success is the same every time the software is executed. How can the probability of success be the same when the software is dependent on its environment, and that environment is continually changing, often in subtle ways? Proponents of software reliability modelling may argue that it is simply a matter of ensuring that the software is exercised in its operational environment. However, it is very difficult to identify all the dependencies of the software on its environment. Failure to do so can lead to misplaced confidence in the software's reliability.

Suppose a software program does not handle 29 February correctly in a leap year. Assuming there were no other defects, had we started running the program on 1 March 2016, then by 28 February 2020 we would have observed 35,040 hours of failure free operation. We would have had a high degree of confidence that the software is correct, yet it would have failed the very next day, 29 February 2020. Software reliability models attempt to predict future software behaviour based on observations of past software behaviour. There is a saying, attributed to Niels Bohr but which is apparently an old Danish proverb, that, "it is difficult to predict, especially the future".

The Ariane 5 accident showed that even a small change in operational profile (from Ariane 4 to Ariane 5) can have an unexpected and catastrophic impact on software behaviour (O'Halloran 2005).

### *3.3.4 Example 3: Defect Distribution*

How are defects distributed by component in software systems? (Hopkins and Hatton 2019) analysed the software defects in a numerical library written in Fortran. They found very strong evidence of defect clustering with typically 80% of all components in the library exhibiting no defect. Furthermore, software errors are more likely to occur at parameter limits and boundaries, which is why software testing techniques such as boundary value analysis are so effective.

Defect clustering is in fact so ubiquitous that it appeared 4$^{th}$ in the "top 10 defect reduction list" compiled by (Boehm and Basili 2001). It follows that if a defect is observed in the implementation of a software feature, it is likely there will be further defects in features implemented by the same software component. Defect clustering turns out to be the result of combining the assumption of uniform distribution of defects across lines of code (or more accurately tokens) with the asymptotic power-law distribution of component lengths which holds for all software systems (Hatton 2014). Suppose then we have a software application that contains ten software components, and that we are told that this software application contains ten defects. The most likely scenario given our prior knowledge is that typically 8 of the 10 components are defect free and the remaining 2 contain all 10 defects. Of course, we don't know which but, as soon as we encounter a defect in a component, we know it's likely to have more defects and we should change our strategy accordingly. Accommodating this in software temporal failure models is by no means obvious, and has not to our knowledge been accomplished.

### *3.3.5 Example 4: Non-Operational Modes and Easter Eggs*

Some software programs have different modes of operation. As well as the normal operational mode, there may be non-operational modes, such as a bootstrap mode, a software update mode, or a debug mode. Should the software inadvertently enter one of these non-operational modes, then it will stop working as expected.

As described in (Ladkin and Littlewood 2016), another variation is that some software contains what are known as Easter Eggs. These are features that are intended to surprise and delight the user. For example, both Microsoft® Excel 97 and Google Earth contain a flight simulator. The Excel 97 flight simulator is launched by creating a new worksheet, pressing F5, typing "L97:X97", pressing Enter, pressing Tab once, holding down Ctrl + Shift and left clicking the Chart Wizard toolbar icon. The Google Earth flight simulator is launched by typing Ctrl + Alt + A. If an Easter Egg is activated inadvertently, then again, the software will stop working as expected.

Easter Eggs is a term also used in the cyber security context. They are "surprises" that can be triggered by environmental conditions decided upon by a malicious actor. Should those conditions be met, the Easter Egg will execute. These conditions may not have been thought of by developers as unintentional conditions that trigger the surprise and, of course, would have been missed by test. Consequently, the software will work normally, possibly for many years, until the right conditions are met, and this evades statistical modelling.

### *3.3.6 Example 5: Duration of the Experiment*

From a statistical point of view, it does not matter whether we run one copy of the software for a million hours or a million copies of the software for an hour each; both experiments result in one million hours of operation. In the real world, these two experiments are not the same. Running a million copies of the software for an hour each does not give us confidence that the software will run for two hours, let alone for days, months or years. Why do we expect the environment's behaviour for one hour to be the same as the environment's continuous behaviour for a million hours?

## 3.4 Limitations

There are some severe limitations that apply even if we accept that the assumptions are valid for a particular software application:

1. A very large number of hours of operation is required to produce a statistically significant result (Butler and Finelli 1993, Kalra and Paddock 2016).
2. Furthermore, the mathematical models used require that we observe no failures. If we observe a failure, we need to start the experiment again. Running a software program for 10,000 hours and observing no failures is not the same as running a software program for 10,000 hours, observing 10 failures and fixing the defects that caused those 10 failures. In both cases, there are no known defects at the end of the experiment. However, in the first case, we would have a high degree of confidence that if we ran the software for another 10,000 hours, we would still see no defects. In the second case, if we ran the software for

another 10,000 hours, we would expect to see further defects (they would just not be the same defects).
3. No software changes can take place during the experiment. If the software is changed in any way, the experiment must be restarted unless justification can be provided as to why the changes made do not invalidate the data collected up to that point. Indeed, this is the approach used in hardware reliability experiments such as those used for airframe stress tests.
4. The software reliability claim is only valid for the software version that was tested, running on the same hardware that was used in the test. If a new software version is released, the experiment must be repeated.
5. The operational profile must be identical.
6. There needs to be an effective system for observing, recording, and documenting faults.

## 3.5 Dissimilar Software

There is a belief that dissimilar software development will defend against common mode or common cause development errors in safety critical systems. In an experiment reported in (Knight and Leveson 1986), the programs were individually extremely reliable but that the number of tests in which more than one program failed was substantially more than expected. (Knight and Leveson 1986) showed that there were common mode errors introduced into software regardless of independence of people, hence diversity didn't achieve the desired outcome. These were common mode errors in the software implementation. Dissimilar software development would not have been able to protect against common mode errors in the software requirements, since the same software requirements were given to all the participants in the experiment.

In a later paper, (Knight and Leveson 1990) stated, "*Until N-version programming has been shown to achieve ultra-high reliability and/or has been shown to achieve higher reliability than alternative ways of building software, the claims that it does so should be considered unproven hypotheses. Until these hypotheses are shown to hold for controlled experiments, depending on N-version programming in real systems to achieve ultra-high reliability where people's lives are at risk seems to us to raise important ethical and moral questions. Attacking us or our papers will not change this*".

On the other hand, (Hatton 1997) concluded, "*The balance of data tends to suggest that N version techniques are preferable in software development when the cost of failure is high due to our inability to make one really good version whatever techniques we currently use. Even though the advantage is much less than that for N independent channels, the difference is still substantial and in this case gave a factor of 5–9 improvement on average for a majority voted 3 version system compared with a single version typical of the current state of the art*". However, this claim was based only on the data from the Knight and Leveson experiment and on the state of the art at the time. More work like (Hatton 1997) is needed in order to widen the samples, and to account for modern development and verification practices.

EASA is promoting dissimilar software. There is a difference of opinion between the FAA and EASA, so dissimilar software is currently the topic of a harmonization effort between the FAA and EASA. EASA have been keen to emphasise that they do not mandate dissimilar software development, but that they do require independence. It is worth examining in this paper, the basis for EASA's position.

At a meeting at which one of the authors was present, EASA stated that for flight controls, EASA will not accept full reliance on Development Assurance and Quality Assurance as sole mitigation of a common mode leading to a total loss of flight controls. When asked, the EASA presenter stated this EASA position is documented in the EASA Generic Certification Review Item (CRI) on "Consideration of Common Mode Failures and Errors in Flight Control Functions" (EASA 2014). This Generic CRI states that the EASA position is that Development Assurance alone is not necessarily sufficient to establish an acceptable level of safety for Flight Control functions, as described specifically in CAST 24 (CAST 2006), and that mitigation means or techniques should be provided to protect against Common Mode Failures/Errors, including software development errors. (EASA 2014) does not specifically address dissimilar software development. (EASA 2014) is a draft CRI that has never been formally issued.

(CAST 2006) does discuss design diversity, dissimilar software, and N-version design. (CAST 2006) cites three papers on design diversity and N-version design, which are (Littlewood 1996), (Littlewood et al. 2000), and (Littlewood et al. 2001). (CAST 2006) has now been withdrawn. It is worth examining here, therefore, what these papers say:

1. (Littlewood 1996) concludes, "*In real applications of this work, it will be necessary to estimate what has actually been achieved, rather than to rely upon the more general results presented here. It is here that hardware engineers have a considerable advantage over their software counterparts. As long as there is sufficient failure data, from previous use of the different component types, that 'covers' the environments within which the new system will operate, the distributions that are needed to compute system reliability will be estimable. This contrasts with the software diversity situation, where estimability is severely constrained by the practical limitations to the number of versions that can be developed*".
2. (Littlewood et al. 2000) claims that, "*The idea that diversity may be a more cost effective way to deliver high diversity is alive and recently it was spelled out by Hatton*" (Hatton 1997). (Littlewood et al. 2000) concludes, "*it appears that Hatton's suggestion that design diversity is always going to be more cost effective than developing a single version software is not trustworthy. In order for us to be certain that diversity will bring more than it takes we need to measure the dependence, which is currently an open question*". We should note that Hatton did not actually suggest that design diversity is "always" going to be more cost effective. Hatton's claim was considerably more circumspect than claimed in (Littlewood et al. 2000) and, as noted earlier, it is recommended that more work like the (Hatton 1997) study is conducted.
3. (Littlewood et al. 2001) presents a theoretical model of hardware redundancy. This model assumes that the presence or absence of each fault (which is claimed to be a random event) is independent of the presence of any other fault, and of the presence in another version. They admit this is the novel assumption that allows them to predict distributions of the probability of failure per demand (pfd) of versions and systems. There is no basis for this assumption, even for hardware, other than that it is a convenience that makes the mathematics tractable. The assumption is certainly not valid for software, as is acknowledged as (Littlewood et al. 2001) concludes, "*However, we do not hide that many of these results are only useful for better understanding these complex, counter-intuitive problems: they do not lead to simple, general recipes for design and assessment. Such understanding is necessary before it is possible to begin engineering diverse fault-tolerant systems with dependability assurance founded on formal models*" and "*Difficulties remain in using some of this theory — most particularly in populating the models with estimates of their key parameters when dealing with real systems*".

In summary, the EASA position is based upon an unissued CRI, which itself was based upon CAST-24 (now withdrawn), which cites the three papers discussed above, which do not appear to support the EASA position.

### 3.6 Are the Models Correct? Are they Useful?

The proponents of software reliability claim that the advantage of assuming that software execution is a Bernoulli process is that the pertinent mathematics is simple, clear, and well understood. This may be so, but it does not follow that a Bernoulli process is an accurate (or even useful) model of software failures.

The use of a Bernoulli process to model software failures is fundamentally flawed. Treating software execution as a Bernoulli process assumes that the probability of success is the same every time the software is executed, which we have shown to be an unwarranted assumption.

This assumption *may* be warranted in specific circumstances. For example, the assumption might be warranted for a software interlock implemented in predicate logic with no state. Where the assumption is unwarranted, it leads to an exaggerated confidence in probabilistic testing and product service history.

### 3.7 Can the Models be Fixed?

(Hatton 2012) presented a mathematical model that hypothesises that software defects follow a power law distribution. This hypothesis was validated by an experiment using 55 million lines of source code. Just as improved financial models have been constructed that use power law distributions, it may be possible to create improved models of software defects using power law distributions. However, defect density is a static property of the software, while failure rates are the result of interaction between the software and its environment over time. While it might be possible to predict the probability of defects using improved models based on Hatton's work, that is not the same as predicting the probability of system failure as a result of the use of that software. If the conditions that trigger the defect are not present then the software will work, and the system will not fail; the converse is also true.

## 4 The Way Forward

### 4.1 Aerospace

(Littlewood and Strigini 1993) concluded that, "*The main point of our paper has been to try to show that here lies the main difficulty when we want to build a system with an ultra-high dependability requirement. Even if, by the adoption of the best design practices, we were to succeed in achieving such a dependability goal, it is our view that we would not know of our achievement at least until the product proved itself after a very great deal of operational use. Nor can we expect to overcome this problem with improved mathematical and statistical techniques*". (Littlewood and Strigini 1993) went on to state, "*Is there a way out of this impasse? One approach would be to acknowledge that it is not possible to confirm that a sufficiently high, numerically expressed dependability had been achieved, and instead make a decision about whether the delivered system is 'good enough' on different grounds. This is the approach currently adopted for the certification of software in safety-critical avionics for civil airliners*". (Littlewood and Strigini 1993) claims, "*this falls far short of providing the assurance needed*".

RTCA/DO-178B (RTCA 1992) was published in 1992. It was superseded by RTCA/DO-178C (RTCA 2011), which made minimal changes to the core document, in 2011. There are now 27,012 commercial jet airplanes in service worldwide (Boeing 2021). No hull loss accidents in passenger service have been ascribed to failure of software developed to RTCA/DO-178B to implement its requirements correctly (Daniels 2011). Software has been a contributing factor in a small number of accidents and in-flight upsets. Modern aircraft and their software are extraordinarily safe (Rushby 2012). We now have nearly 30 years of service experience that satisfying the objectives of RTCA/DO-178B/C has been sufficient to address the software considerations in aircraft certification.

### 4.2 Proof and Economics

We contend that our ability to make one really good version has improved in the last 30 years through the introduction of improved software development and verification techniques. In particular, the processing power of modern CPUs has made automated program proof increasingly tractable. Formal methods allow us to verify that a software artefact is correct, complete, and unambiguous.

There are plenty of anecdotes that claim that formal methods will save developers time and money; some may even say that it is 'better' than any other approach. Indeed, (Rushby 2009) and (Littlewood and Rushby 2012) wrote that formal methods may have a part to play in providing higher assurance based upon the idea introduced by (Littlewood 1998) that software may be "possibly perfect". The general proposition is that, "possible perfection provides a bridge between the verification activities used to ensure correctness of software and the probabilistic estimates required for failure at the system level". Given that use of formal methods can qualitatively achieve 'at least as good software' as any other approach, the question is what evidence is there that it does save time and money?

It has been the case for a number of years that large developers, such as Google, Facebook, Amazon, and Microsoft have been recruiting formal methods expertise with the aim of being more robust (Garavel et al 2020). For example, (Ball et al. 2004) describes how Microsoft Research developed a Static Driver Verifier using formal methods. This tool is now widely used to verify third party Windows device drivers and has greatly reduced the number of Blue Screens of Death (BSOD). Again, this is a qualitative argument, but the market appears to have its view. While there exists the possibility that better software reliability models could be developed, the only techniques available at the present time that can predict accurately the behaviour of software before it is executed are formal methods. An approach therefore might be to use formal methods wherever practicable and to back this up with testing and/or simulation where needed in order to increase confidence.

The use of formal techniques, like any software development technique, has to be based upon assumptions about the environment and there are limits to what these techniques can, or indeed should be, used for (Murray and van Oorschot 2018).

### 4.3 An Industrial Experiment

An experiment, run by Warwick University Manufacturing Group in 2018, on automotive software showed the benefits of the use of formal techniques. The blind trial involved benchmarks of real automotive requirements and design for 6 functions that had been seeded with 48 errors. The trial measurement process was developed by York Metrics and time taken for each stage of the verification process was recorded. There were 3 processes used independently by 3 different people: baseline human review, tool assisted using Simulink Design Verifier (SLDV) and a formal method-based tool developed by D-RisQ Ltd called Modelworks®, and in all 3 cases the 48 errors were detected. In the case of Modelworks®, a 49th error was discovered that had not been seeded. However, the time for Modelworks® to discover these errors was between 60–80% faster across all 6 cases measured, in one case reducing from 52 hours down to 10 hours; see Figure 1. It will be noted that there are 7 sets of measurements. Experiment PP failed to complete analysis as time for the project ran out, though Modelworks® found the relevant errors. The 7th case is a repeat of TA (i.e., TA2) by a different set of 3 people in a second company applying their own processes for the benchmark. It can be seen that the results are very close (but not the same) thus demonstrating consistency in the experimental results.

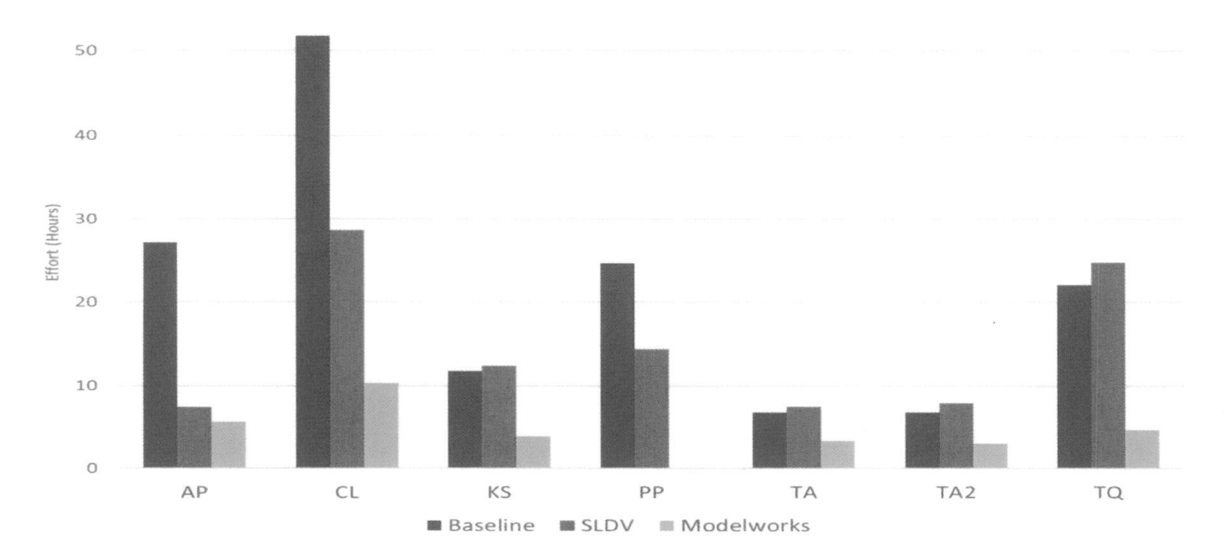


**Figure 1 ~ Industrial Scale Benchmarking Study**

### 4.4 Requirements Engineering

The weak link, for avionics software at least, remains requirements engineering. There has not been a single hull loss accident in passenger service that has been ascribed to failure of software developed to RTCA/DO-178B/C to implement its requirements correctly. There have been several accidents where software implemented the requirements correctly, but those requirements specified unsafe behaviour under some unforeseen circumstance. For example, an Airbus A320 overran the runway at Warsaw on 14 September 1993 (MCAAI 1994). A contributing factor was that deployment of the ground spoilers and engine thrust reversers was delayed because of a requirement to deploy them only when both main landing gear struts indicated Weight on Wheels. In another accident, an Airbus A320 overran the runway at Sao Paulo on 17 July 2007 (CENIPA 2009). A contributing factor was that the pilot only pulled one thrust lever into the reverse thrust position (the other thrust reverser was known to be inoperative), but a requirement stated that both thrust levers must be in the idle or reverse thrust position for either of the thrust reversers to be deployed. Finally, in the two recent Boeing 737 MAX accidents on 29 October 2018 and 10 March 2019, the Manoeuvring Characteristics Augmentation System (MCAS) software implemented its requirements correctly, but the requirements caused full nose down trim to be applied following an Angle of Attack sensor failure (Daniels 2020).

As Nancy Leveson has said, "Software-related accidents are usually caused by flawed requirements". It therefore follows that our efforts should be focused on writing better requirements. Formal methods can help with writing better requirements by using formal requirements languages with unambiguous semantics and formal methods tools that can ensure the requirements are complete and consistent.

## 5 Conclusion

We have shown in this paper that software reliability models that assume that software execution is a Bernoulli process are flawed. The assumption of independence may be warranted in specific circumstances, but otherwise these models lead to an exaggerated confidence in probabilistic testing and product service history. These models cannot, in general, be used to demonstrate ultra-high reliability.

There seems to be an opportunity for the research community to investigate further the efficacy of N-version programming in mitigating the impact of software errors.

The situation is not as unsatisfactory as was stated in (Littlewood and Strigini 1993). We have now accumulated nearly 30 years of service experience that satisfying the objectives of RTCA/DO-178B/C has been sufficient to address the software considerations in aircraft certification.

The state of the art has improved considerably since RTCA/DO-178B was published in 1992. Formal methods are now used more widely, and automated program proof is now tractable and cost-effective.

Requirements continue to be the weak link. Analysis of historical aircraft accidents suggests that in those accidents where software was involved, the software implemented its requirements correctly, but the requirements specified behaviour that was unsafe and led to the accident.

We need to focus on getting the requirements right if we are to improve safety.

**References**

Ball, T., Cook, B., Levin, V. & Rajamani, S.K. (2004). *SLAM and Static Driver Verifier: Technology Transfer of Formal Methods inside Microsoft* (Technical Report MSR-TR-2004-08). Microsoft Research. Retrieved 22 September 2021 from https://www.microsoft.com/en-us/research/wp-content/uploads/2016/02/tr-2004-08.pdf

Boehm, B. & Basili, V. R. (2001). Software Defect Reduction Top 10 List. In *IEEE Computer. January 2001,* 135–137. Retrieved 8 November 2021 from https://www.cs.umd.edu/projects/SoftEng/ESEG/papers/82.78.pdf

Boeing. (2021). *Statistical Summary of Commercial Jet Airplane Accidents, Worldwide Operations 1959 – 2020*, Boeing. Retrieved 9 November 2021 from https://www.boeing.com/resources/boeingdotcom/company/about_bca/pdf/statsum.pdf

Butler, R.W., & Finelli, G.B. (1993). The infeasibility of quantifying the reliability of life-critical real-time software. In *IEEE Transactions on Software Engineering, 19(1),* 3–12. Retrieved 28 September 2021 from https://shemesh.larc.nasa.gov/fm/papers/Butler-nonq-paper.pdf

CAST. (2006). *Reliance on Development Assurance Alone When Performing a Complex and Full-Time Critical Function* (Position Paper CAST-24 Rev 2). FAA Certification Authorities Software Team.

CENIPA. (2009). *Final Report A – No 67/CENIPA/2009.* Retrieved 12 November 2021 from http://sistema.cenipa.aer.mil.br/cenipa/paginas/relatorios/rf/en/3054ing_2007.pdf

Daniels, D. (2011). *Thoughts from the DO-178C committee.* Paper presented at the 6th IET International Conference on System Safety, 2011.

Daniels, D. (2020). *The Boeing 737 MAX Accidents.* Paper presented at the 28th Safety-Critical Systems Symposium, York, UK. Retrieved 9 November 2021 from https://scsc.uk/rp154.1:1

EASA. (2014). *Consideration of Common Mode Failures and Errors in Flight Control Functions* (Draft Generic CRI D-05), European Aviation Safety Agency.

Garavel, H., ter Beek. M. & van de Pol, J. (2020). *The 2020 Expert Survey on Formal Methods.* Paper presented at the 25th International Conference on Formal Methods for Industrial Critical Systems, Vienna, Austria. Retrieved 9 November 2021 from https://hal.inria.fr/hal-03082818/document

Hatton, L. (1997). *Are N average software versions better than 1 good version?* IEEE Software, SE-14(6), 71–76. Retrieved 22 September 2021 from https://www.leshatton.org/Documents/Nver_1297.pdf

Hatton, L. (2012). *Power-laws and the Conservation of Information in discrete token systems: Part 2 The role of defect.* Retrieved 17 September 2021, from https://arxiv.org/abs/1209.1652

Hatton, L. (2014). *Conservation of Information: Software's Hidden Clockwork?* IEEE Transactions on Software Engineering, 40(5), 450–460. Retrieved 8 November 2021, from https://www.leshatton.org/Documents/TSE-2013-08-0271_V1.0a.pdf

Hopkins, T.R. & Hatton, L. (2019). *Defect patterns and software metric correlations in a mature ubiquitous system.* Retrieved 17 September 2021, from https://arxiv.org/abs/1912.04014

IEC. (2010a). *Functional safety of electrical/electronic/programmable electronic safety-related systems — Part 3: Software requirements.* International Electrotechnical Commission.

IEC. (2010b). *Functional safety of electrical/electronic/programmable electronic safety-related systems — Part 7: Overview of techniques and measures.* International Electrotechnical Commission.

Kalra, N., Paddock, S.M. (2016). *Driving to Safety: How Many Miles of Driving Would It Take to Demonstrate Autonomous Vehicle Reliability?* RAND Corporation. Retrieved 11 November 2021 from https://www.rand.org/pubs/research_reports/RR1478.html

Knight, J.C. & Leveson N.G. (1986). *An experimental evaluation of the assumption of independence in multi-version programming.* IEEE Transactions on Software Engineering, 12(1), 96–109. Retrieved 22 September 2021 from http://sunnyday.mit.edu/papers/nver-tse.pdf

Knight, J.C. & Leveson N.G. (1990). *A reply to the criticisms of the Knight & Leveson experiment.* ACM SIGSOFT Software Engineering Notes, *15*(1), 24–35. Retrieved 8 November 2021 from http://sunnyday.mit.edu/critics.pdf

Ladkin, P.B. & Littlewood, B. (2016, February). *Practical Statistical Evaluation of Critical Software.* Paper presented at the 24th Safety-Critical Systems Symposium, Brighton, UK. Retrieved 27 September 2021 from https://scsc.uk/r131/8:1

Littlewood, B. & Verrall, J.L. (1973, April). *A Bayesian reliability growth model for computer software.* Paper presented at the IEEE Symposium on Computer Software Reliability, New York, USA.

Littlewood, B. & Strigini, L. (1993). *Validation of Ultrahigh Dependability for Software-Based Systems.* Communications of the ACM (CACM), 36(11), pp. 69–80. Retrieved 9 November 2021 from https://openaccess.city.ac.uk/id/eprint/1251/1/CACMnov93.pdf

Littlewood, B. (1996). *The impact of diversity upon common mode failures*. Reliability Engineering & System Safety, 51(1), 101–113. Retrieved 22 September 2021 from https://openaccess.city.ac.uk/id/eprint/1630/

Littlewood, B. (1998) *The use of proof in diversity arguments*. IEEE Transactions on Software Engineering, 26(10), 1022–1023. https://www.researchgate.net/publication/3188113_The_Use_of_Proof_in_Diversity_Arguments_in

Littlewood, B., Popov, P. & Strigini, L. (2000). *N-version design versus one good version.* Paper presented at the International Conference on Dependable Systems & Networks, New York, USA. Retrieved 22 September 2021 from https://www.researchgate.net/publication/2585976_N-version_design_Versus_one_Good_Version

Littlewood, B., Popov, P. & Strigini, L. (2001, February). *Design Diversity: an update from research on reliability modelling.* Presented at the 9th Safety-Critical Systems Symposium, Bristol, UK. Retrieved 22 September 2021 from https://openaccess.city.ac.uk/id/eprint/261/2/SCSS2001_v_12.forDistrib.pdf

Littlewood, B. and Rushby, J. (2012). *Reasoning about the Reliability of Diverse Two-Channel Systems in Which One Channel is "Possibly Perfect".* IEEE Transactions on Software Engineering, 38(5), 1178–1194. Retrieved 9 November 2021 from https://openaccess.city.ac.uk/id/eprint/1069/1/1oo2-revised-13apr11.pdf

Mandelbrot, B.B & Hudson, R.L. (2004). *The (Mis)Behavior of Markets: A Fractal View of Risk, Ruin and Reward,* Basic Books.

MCAAI. (1994). *Report on the Accident to Airbus A320-211 Aircraft in Warsaw on 14 September 1993*, Main Commission Aircraft Accident Investigation Warsaw, March 1994. Retrieved 9 November 2021 from http://www.rvs.uni-bielefeld.de/publications/Incidents/DOCS/ComAndRep/Warsaw/warsaw-report.html

Murray, T. and van Oorschot, P. C. (2018). *BP: Formal Proofs, the Fine Print and Side Effects.* Paper presented at 2018 IEEE Cybersecurity Development (SecDev). Retrieved 9 November 2021 from https://people.scs.carleton.ca/~paulv/papers/secdev2018.pdf

O'Halloran, C. (2005). *Ariane 5: Learning from Failure.* Proceedings of the 23rd International System Safety Conference, San Diego, 2005.

RTCA. (1992). *Software Considerations in Airborne Systems and Equipment Certification*, RTCA/DO-178B, RTCA, Inc. Also available as EUROCAE Document ED-12B.

RTCA. (2011). *Software Considerations in Airborne Systems and Equipment Certification*, RTCA/DO-178C, RTCA, Inc. Also available as EUROCAE Document ED-12C.

RTCA. (2021). *Terms of Reference, Special Committee (SC) 240, Topics on Software Advancement (Revision 1)*, RTCA Paper No. 083-21/PMC-2139, RTCA, Inc. retrieved 9 November 2021 from https://www.rtca.org/wp-content/uploads/2021/05/SC-240-TOR-Rev-1-Approved-2021-03-18.pdf

Rushby, J. (2009). *Software Verification and System Assurance*. Presented at the 7th IEEE International Conference on Software Engineering and Formal Methods (SEFM), Hanoi, Vietnam, November 2009. In SEFM Proceedings pp. 3–10. Retrieved 22 September 2021 from http://www.csl.sri.com/users/rushby/papers/sefm09.pdf

Rushby, J. (2012). *New Challenges in Certification for Aircraft Software.* Presented at the Ninth ACM International Conference on Embedded Software (EMSOFT), Taipei, Taiwan. Retrieved 22 September 2021 from http://www.csl.sri.com/users/rushby/papers/emsoft11.html

This collation page left blank intentionally.

ISSN 2754-1118 (Online) — ISSN 2753-6599 (Print)

# ‘til the Next Zero-Day Comes

### Ransomware, Countermeasures, and the Risks They Pose to Safety

**Bruce Hunter**

Safety and Security Consultant, Sydney, Australia

## Abstract

*Cyber-attacks on critical infrastructure are not new, but their recent intensity has increased the risk of intended or unintended consequences to safety systems to become a real and present danger. Ransom use of malware attacks have mainly concentrated on business systems, by denying access to essential data, but recent attacks have affected critical infrastructure with consequential shutdown of operation-al technology including safety-related functions. Although ransomware may intentionally cause dangerous failures in the system, pervasive connectivity raises the risks of this happening. This article discusses the precursors to this danger as part of Information Technology and Operational Technology convergence, integration of business and control systems, conflicts arising out of this integration and monetarisation of vulnerability exploitation. Although using Industrial Control System examples are used, safety practitioners may use these to mitigate cybersecurity threats and minimise the impact of attacks on all safety-related systems and their recovery.*

## 1 Out of Gas!

Usually, cyber-attacks on critical infrastructure that includes safety-related systems go relatively unnoticed, unless they have a significant impact on the public. Partly this is due to the sensitivity of companies to bad press and cautiousness in trying to prevent copy-cat attacks. An understanding of the magnitude and nature of attacks, however, can be gained from published, but anonymized, surveys of the industries concerned.

Ransomware attacks in general have become a major threat. Industry surveys have also shown that ransomware is the cyber-threat most likely to affect their organization in the next 12 months (ISACA 2021):

- 21% of businesses in general said that their organizations have experienced ransomware attacks
- 67% said that their organizations will take new precautions in light of the attack on Colonial Pipeline, an American oil distribution system from Houston, Texas
- 78% said critical infrastructure organizations should *not* pay ransom if attacked
- 84% said ransomware attacks would become more prevalent in the second half of 2021

A 2020 survey (Bakuei et al. 2021) shows, for the US alone, that 19 organisations found ransomware in their Industrial Control Systems (ICS).

Ransomware risk to Operational Technology (OT) including ICS differs from other forms of attack in the following ways:

- The primary objective of ransomware is monetary gain by locking up assets and information needed by the target organisation
- Impact on critical functions including safety is generally unintended and consequential when resources needed by these functions become unavailable
- Ransomware is only designed to interrupt and may contain untested side effects, which can have reliability and safety implications. In the Colonial Pipeline case, the de-encryption tool did not work properly, and recovery required other methods.

The ransomware attack on the US Colonial Pipeline company in 2021 certainly did get noticed and led to the shutdown of the gas distribution operation for a week affecting the supply of over 40% of the US East Coast supply of automotive and aviation fuel. Like most complex failures, this resulted from a series of events and decisions that made the company's systems more vulnerable, and the response to the attack less optimal.

**Figure 1 ~ "Colonial Pipeline", Photograph by Orbital Joe**

The key events of this attack were (Charles Carmakal, 2021):

- Hackers gained entry into the networks of Colonial Pipeline on April 29 through a forgotten virtual private network (VPN) account that had poor security (Attributed a cybercrime group using to DarkSide Ransomware-as-a-service (RAAS) toolkit)
- Just before 5am on May 7, Colonial's control room saw a ransom note on their system
- DarkSide ransomware attack locked out the business operation
- As a safety precaution, Colonial shut down its entire pipeline operation, causing critical fuel shortage to Eastern US
- After extensive checks of the OT network concluded that damage was limited to some of the Information Technology (IT) business operation, supply restarted on May 13
- Colonial did pay USD 4.4M in Bitcoin ransom, as a precaution in case recovery was not achievable, although the Federal Bureau of Investigation (FBI) was subsequently able to recover most of this ransom

The following precursors the industry faced in the last few years may have strongly influenced the reaction to this event:

- The US Department of Homeland Security (DHS) Cybersecurity & Infrastructure Security Agency (CISA), and the FBI, advised Pipeline Operators of a spear-phishing and intrusion campaign conducted by state-sponsored Chinese actors that occurred from December 2011 to 2013, targeting U.S. oil and natural gas pipeline companies (CISA and FBI 2021a).
- A US Congressional report highlights pipeline cybersecurity issues, but US Transportation Security Administration (TSA) "*maintains that voluntary cybersecurity standards have been effective in protecting US pipelines from cyber-attacks*" (Parfomak, 2012)

- The United States Government Accountability Office conducted a review of "*TSA's efforts to assess and enhance pipeline security and cybersecurity*" and issued 10 improvement recommendations in its congressional report, GAO-19-48, but did not highlight industry cybersecurity regulation (GAO 2018).
- In 2019 alone, there were 614 reported pipeline incidents in the United States, resulting in the death of 10 people, injuries to another 35, and about $259 million in damages (Kelso, 2020).
- Despite warnings of pipeline specific ransomware threats (CISA 2020), the industry has been criticised for poor cybersecurity practices and lobbying against stronger regulation
- Industrial Control Systems (ICS) have become an attractive target to Ransomware attackers. The motives are strong (money), the risk is low (RAAS toolkits protect attacker), the target is soft (old ICS hidden vulnerabilities abound), and reward is fairly certain (ICS operations are critical). (Palmer, 2021)
- The Colonial Pipeline company supplies about 45% of east US coast fuel, making it a major risk to US transport operations
- Poor security practices would have meant uncertainty on the reliability of OT network segmentation.

These precursors could have influenced the company's response to the ransomware attack and its outcome, which did result in the shutting down the fuel supply.

Colonial's east coast pipeline is a multi-product, multi-offtake, multi-line, multi-section, and multi-storage operation that has high risks associated with operations outside the safe envelope. Out of an abundance of caution to ensure a dangerous situation did not arise, the company chose to shut down operations to prevent spread to the OT network until the business system was cleared of ransomware and put back into operation (Hoffman and Winston 2021).

The system failure resulted in the following timeline of events with the US fuel supply:

- Crisis caused by the loss of supply resulted in the Federal Motor Carrier Safety Administration (FMCSA) declaring a state of emergency in 18 states to help with the shortages (9 May 2021)
- Colonial Pipeline eventually re-established pipeline operation (13 May)
- FBI and CISA issue alerts on pipeline ransomware threat (11 and 19 May)
- TSA update to Pipeline Security Guidelines issued, replacing criticality guidelines - naturally (TSA 2021a)
- TSA issued Security Directive Pipeline-2021-01 (TSA 2021b) directing a whole range of mandatory report and assessments with significant penalties for non-compliance (27 May)
- United States Department of Justice (DOJ) gives critical infrastructure ransomware attacks equivalent priority to terrorism. (3 June)
- CISA, with the FBI, updated a joint advisory on DarkSide Ransomware: "Best Practices for Preventing Business Disruption from Ransomware Attacks" (CISA and FBI 2021b), originally published 11 May, to supplement previous advice (8July).
- On 28 July, the US president announced further steps to safeguard critical infrastructure (Biden 2021)

Consequences of the Colonial Pipeline attack have certainly focused the minds of the process sector (Moore 2021); but why the concern? Ransomware is a substantial risk to control systems due to:

- Difficulty to recover operations as systems locked by encrypted files
- High motivation of attackers due to monetary rewards or nation-state intent

- Possibility of accidental damage to safety systems as a by-product, due to unaccountability of cybercriminals compared to nation-state threat actors

Ransomware Lessons Learned for safety from this ransomware attack includes:
- Safety and security must be coordinated. This could have affected safety elements of the pipeline if essential operator controls, e.g. human-machine interfaces (HMIs), were affected (CISA 2020), or if the security response to ransomware had impact on safety.
- OT operational aspects may have continued if, independent of the business systems.
- Segregation between OT and IT is necessary but not always assured.

# 2 Who's in control? OT Cybersecurity

## 2.1 An Analogy

As I write this article in mid-2021, Sydney is in the middle of another lockdown, trying to control a COVID-19 Delta-strain outbreak despite months when there were effectively no cases in the country. Pandemics and the way we deal with them are very reminiscent of cybersecurity. The following issues are very much characteristic of pandemic and cybersecurity experiences:
- Failure consequences can be far-reaching and include morbidity
- Responses and their success are driven by
  - risk appetite of participants
  - motivation of the threat
- Protection involves:
  - separation of the vulnerable from the threat
  - continually increasing the resilience of the target
  - educating the user on responsibility and actions that may increase risk
- but Thwarted by:
  - constant evolution of the threat tactics and vulnerable entry points
  - detection reliant on known indications and subject to false positives and negatives
  - lack of persistence in assessing and addressing the risks
  - the risk that defence may actually harm the defended
  - unpopularity of preventive measures
  - cognitive bias and complacency about risk

## 2.2 Convergence of Technology

Control systems have evolved from mechanical, electrical, electronic, and programmable electronic systems but usually remained standalone. OT has converged with hardware and software of IT with the benefits of economies of scale, supportability of common operating systems and skillsets. A less welcome consequence of this was the inheritance of visibility and vulnerability to cyber-attack (ISACA 2016).

Adopting IT platforms and Operating Systems specifically provides:
- Economies of scale;
  - but increased vulnerabilities with complex designs, update frequency and obsolescence issues
- Commonality of software and hardware providing a wealth of functionality;
  - but inheritance of IT vulnerabilities and attack toolkits

  - o and visibility to IT threat agents
- Increased connectivity to allow expanded operation and monitoring;
  - o but expanded attack vectors or surfaces
- Increased functionality and capability;
  - o but reduced reliability and predictability

These issues have made OT systems not only more vulnerable to cyber-attack, but also have impacted reliability and safety.

## 2.3 ICS Threats

Cyber-threats to systems incorporating Operational Technology (OT) have grown to a point where, in 2020, ICS Advisories were issued by US CISA at an average rate of 5 per week (https://www.cisa.gov/uscert/ics/advisories). Analysis shows that, advisories and alerts have grown to an average of 7 per week in 2021: 55% more than 2020 (Figure 2).

**Figure 2 ~ Analysis of CISA ICS Advisories Over Time**

Threat actors to ICS have evolved from script buddies and hacktivists to now include state-sponsored specialists and cybercrime using ransomware-as-a-service (RAAS) toolkits.

Significant past attacks noted by CISA of non-ransomware cases with ICS include:

- Joint CISA-FBI Cybersecurity Advisory (AA21-201A) (CISA and FBI 2021a): Gas Pipeline Intrusion Campaign. Attributed to China nation-state cyber actors. Of the targeted entities, 13 were confirmed compromises, 3 were near misses, and 7 had an unknown depth of intrusion. Impact is development by the attacker of capabilities against U.S. pipelines to physically damage pipelines or disrupt pipeline operations
- W32.DistTrack, also known as "Shamoon" (CISA 2021a): An information-stealing malware that also includes a destructive module. Attributed to Iranian nation-state cyber actors. Operational impacts of this attack include loss of intellectual property (IP) and disruption of critical systems.
- ICS Focused Malware Havex Trojan (CISA 2018): An information-stealing malware. Attributed to Russian nation-state cyber actors. Operational impacts may result from information gathered in this attack or malware installed.

- Malware campaign that has compromised numerous industrial control systems (ICSs) environments using a variant of the BlackEnergy malware (CISA 2021b): Installs malware on Internet-connected HMIs. Attributed to Russian nation-state cyber actors. Impact is other actions enabled by compromised HMI.
- Cyber-Attack against Ukrainian Critical Infrastructure including electricity grid (CISA 2021c): Attributed to Russian nation-state cyber actors. Impact was the takeover of electricity grid HMI operation, shutdown communication and backup recovery with loss of large section of the grid.
- CrashOverride Malware (CISA 2021d): Attributed to Russian nation-state cyber actors. Impact is the abuse of functionality in a targeted ICS system's legitimate control system to achieve its intended effect which has included shutdown of electricity grid using standard ICS protocols but could impact all critical infrastructure organizations.
- Safety System Targeted Malware HatMan, also known as TRITON and TRISIS (CISA 2019): Attributed to Russian Government-Owned Laboratory but possibly used by Iranian nation-state cyber-actors. Disrupted the safety-related triple redundant Emergency Shutdown (ESD) of Saudi Arabian petrochemical plant. Safety protection key switch (SIS configuration mode) on one channel was left in the program state allowing the modification of the Triconex system; the other two channels detected anomaly of exploited channel and shutdown.

## 2.4 Vulnerabilities and Attack Paths

The vulnerability-patch cycle with converged IT-OT, leads threat agents to go to deeper technology layers to achieve exploitation, as examples show in Figure 3.

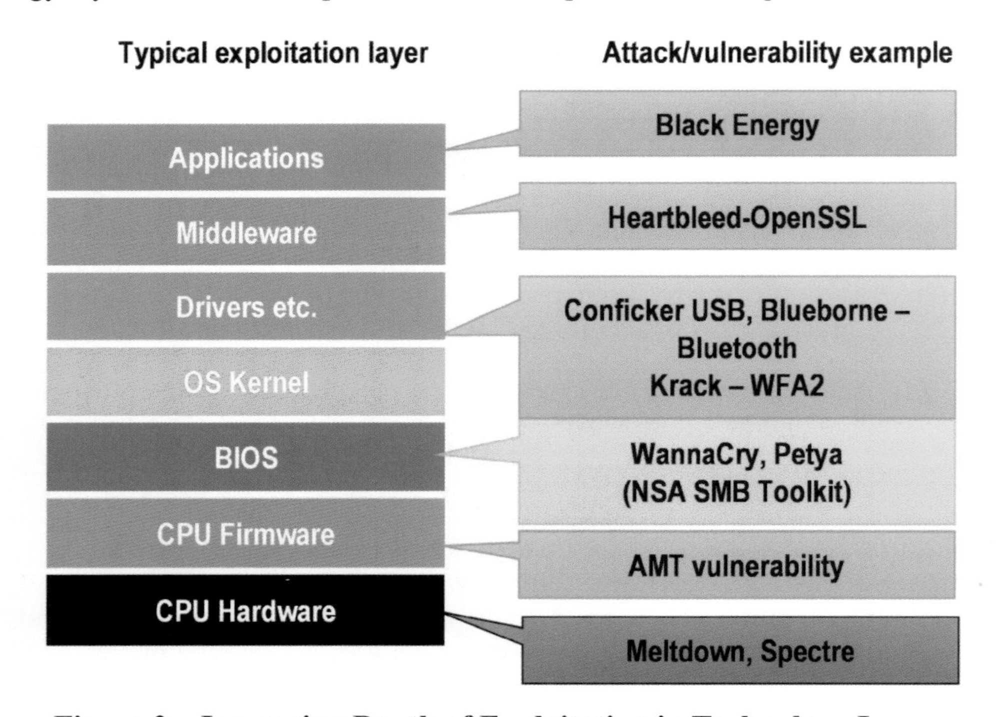


**Figure 3 ~ Increasing Depth of Exploitation in Technology Layers**

Attack paths and surfaces used by adversaries rely on accessibility to ICS OT networks. Understanding the risk of entry and protective measures is helped by a layered architecture as the example in Figure 4 modified from IEC 62443 (IEC n.d.), which is loosely based on Purdue (Purdue 1989).

This layered approach does provide a Defence-in-depth approach but exhibits some key vulnerable attack vectors as experience in ICS attacks:

- Entry via vulnerabilities in the web and email services of the Enterprise layer (Purdue Layer 4). Despite the use of Demilitarised Zone (DMZ) technology there are still backdoors to attacks via techniques such as email spear phishing
- Entry via Operations Control (Purdue Layer 3); DMZ again subject to backdoor entry
- Entry via Supervisory Control (Purdue Layer 2) and Controllers (Purdue Layer 1) via backdoors and malware on HMI and engineering workstation. Safety-related systems at this level have to be protected against dangerous attacks. Entry points at this level include IT technology of HMI and Engineering Workstations, again typically by malware carried on spear-phishing e-mail or remote maintenance network connection. Even if this is precluded by "air-gaps", backdoors can be gained via portable media (e.g. USB Drives) used to update system software or firmware.
- Manipulation of standard protocols in Field Device connections (Purdue Layer 1 and 0).

Network segmentation (restricting data flow) is a key technique in not only providing defence-in-depth against attack but also providing functional separation or independence for different critical elements of an ICS. Network segmentation does not necessarily mean functional separation.

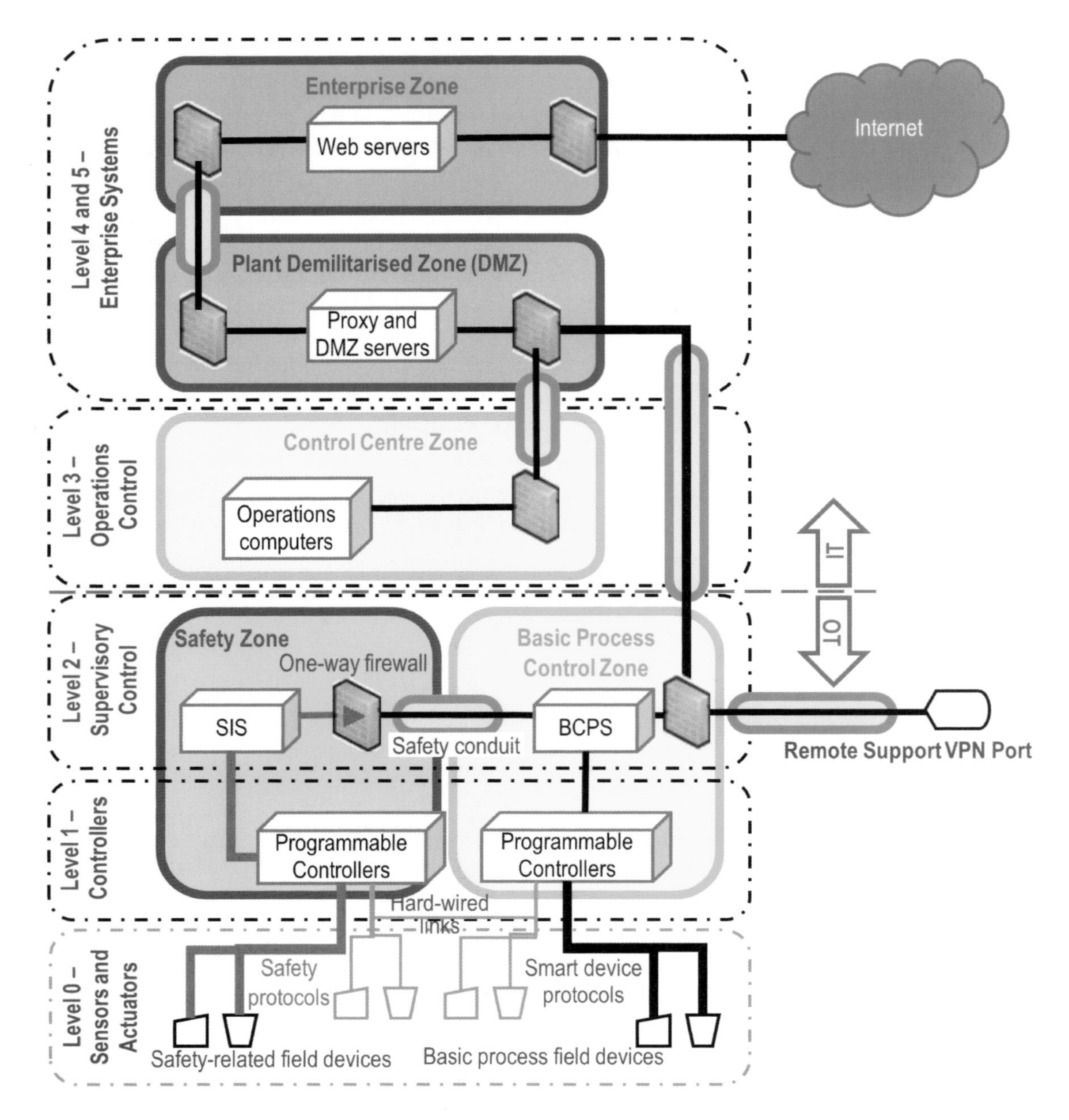


**Figure 4~ Simplified Layered ICS Security Architecture (IEC 62443)**

Evolution of OT architecture to Industrial Internet of Things (IIoT), Cloud Services and Nested Edge Technology is providing emerging challenges to true defence-in-depth, segmentation, and cybersecurity protection of safety-related systems. Protective segmentation becomes difficult in Software Defined Perimeters due to logical architecture as shown in Figure **5**, which is extracted from Annex C of draft IIoT standard ISO/IEC FDIS 30162. Proof of dependable functional separation and interaction is difficult to prove in software-defined networks with non-industrial protocols. Standards are catching up on the model and requirements for IIoT, but issues remain in proving trustworthy segmentation of functional layers. IEC/ISO JTC1/SC41 has highlighted these issues for future Standardization work in a technical report (ISO/IEC 2020).

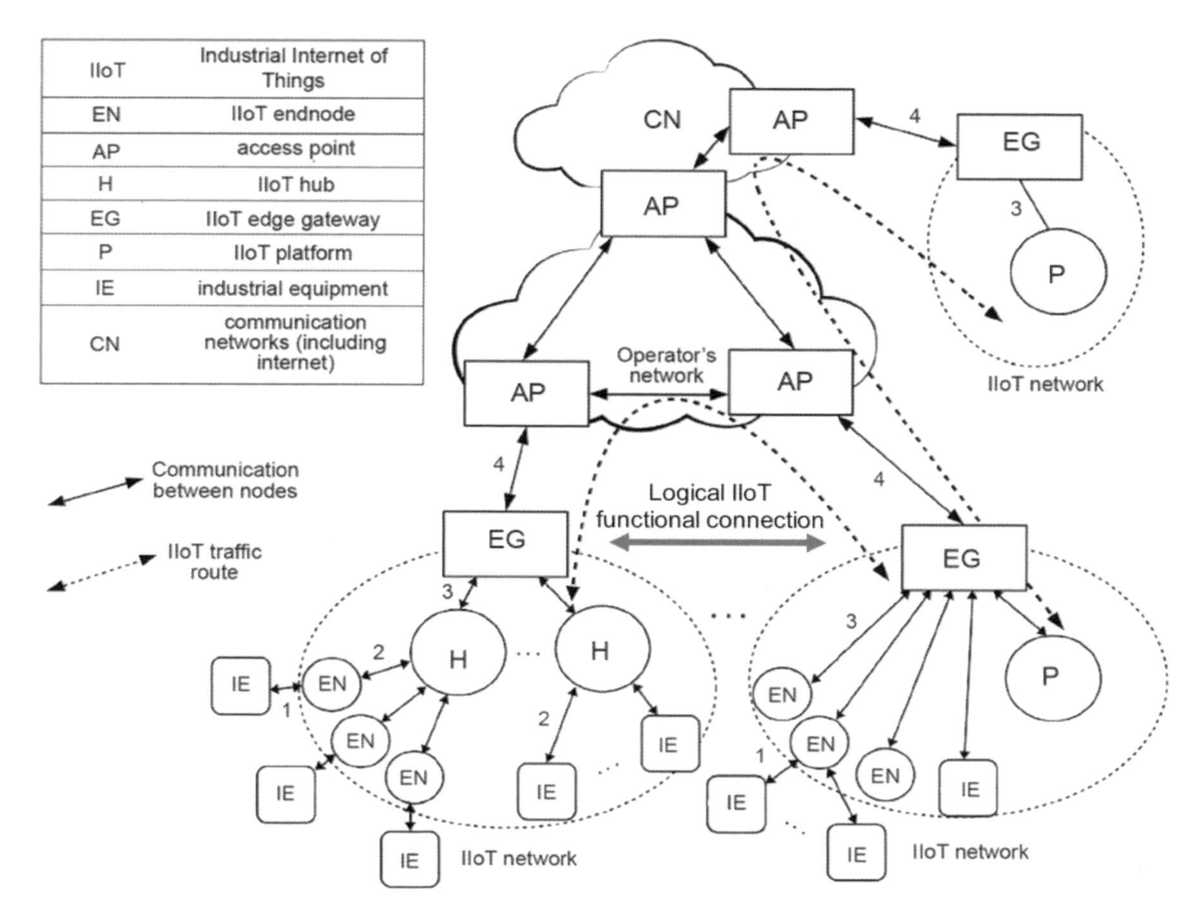


**Figure 5 ~ Logical Segmentation of Network in IIoT-Cloud Architecture**

## 2.5 Standards, Directives and Frameworks

### *2.5.1 General*

Evolving threats to and vulnerabilities of ICS have led to a variety of standards, guides, and framework to protect these systems. There is growing guidance on securing OT in ICS (Timpson and Moradian 2018) and standards to support these.

### *2.5.2 Ransomware Specific Frameworks*

The following specific frameworks have been established to minimise the specific risk of ransomware to ICS including pipeline systems:

- CISA FBI joint advisory on DarkSide Ransomware: Best Practices for Preventing Business Disruption from Ransomware Attacks (AA21-131A) (CISA and FBI 2021b)
- US TSA updated Pipeline Security Guidelines (TSA 2021a)
- NIST Framework with security lifecycle guidance on the application of the NIST CSF to ransomware mitigation (NIST 2021).

### *2.5.3 The ISO/IEC - Standard Security for Industrial Automation and Control Systems*

IEC 62443 (IEC n.d) is the "go-to" standard for OT specific cybersecurity, rather than the ISO/IEC 27000 series commonly used for IT, providing requirements to protect against various levels of cyber-attack. It also supports specific protection of safety-related aspects. The published parts include requirements for developers, asset owners, operators and assessors covering the following fundamentals, as follows:

- Security Governance (SG) – having 42 requirements
- Security Development and Integration (SDI) – having 45 requirements
- Risk Management (RM) – having 74 requirements
- Asset Management (AM) – having 128 requirements
- Identification and Authentication Control (IAC) – having 75 requirements
- Use Control (UC) – having 84 requirements
- System Integrity (SI) (includes integrity of safety functions) – having 84 requirements
- Information Confidentiality (IC) – having 47 requirements
- Restricted Data Flow (RDF) – having 30 requirements
- Incident Management (IM) – having 68 requirements
- Resource Availability (RA) – having 29 requirements

### *2.5.4 NIST Cybersecurity Framework (CSF)*

The NIST Framework for Improving Critical Infrastructure Cybersecurity (NIST 2018a), as illustrated in Figure 6, provides a model for assessing maturity of cybersecurity processes.

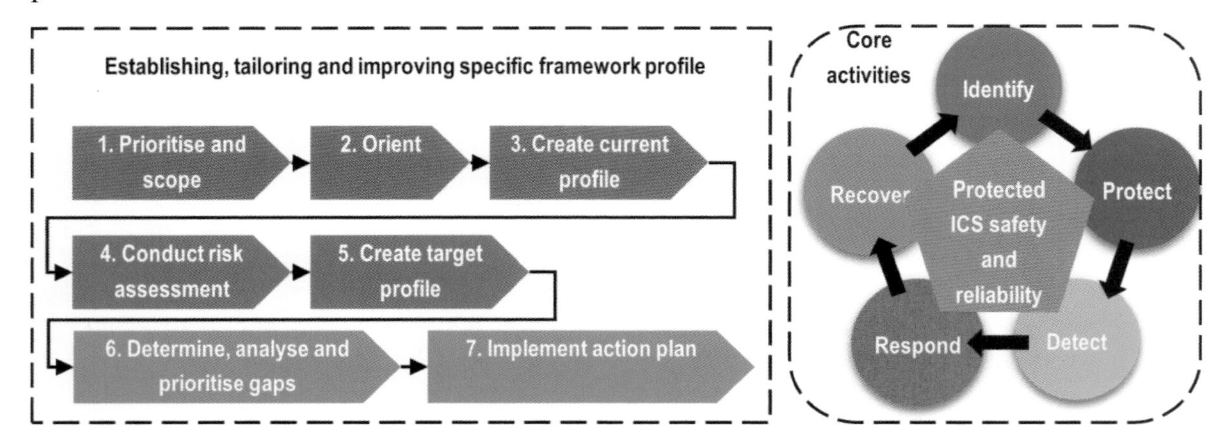


**Figure 6 ~ NIST Cybersecurity Framework Model**

Safety interaction with the NIST CSF as part of the kill chain is illustrated in Figure 10 of Sub-section 3.3.

Appendix A, Section A.2, proposes a tailoring of the framework core for safety-related systems. CSF profiles have been published for various critical infrastructure sectors.

## 2.6 Maintaining Effectiveness of ICS Cybersecurity Protection

### *2.6.1 Patching and Updates*

The effectiveness of ICS cybersecurity countermeasures is only as good as the vulnerabilities and threats address at the time of the last update ('til the next zero-day).

Effective cybersecurity protection relies on prompt patches of system vulnerabilities. This has the following challenges with OT systems:

- Reliability of the control system can be degraded by unproven updates. Reliability growth in OT is built up over time and by correction of systematic errors. Patches may cause regression in this growth.
- Security patches may cause unexpected failure due countermeasures or malware detection patterns (Goodin 2010).

Patches and updates to OT systems require validation before they are put on live ICS (Hunter 2013). Figure 7 illustrates the cycle of new vulnerabilities, patching validation and system changes.

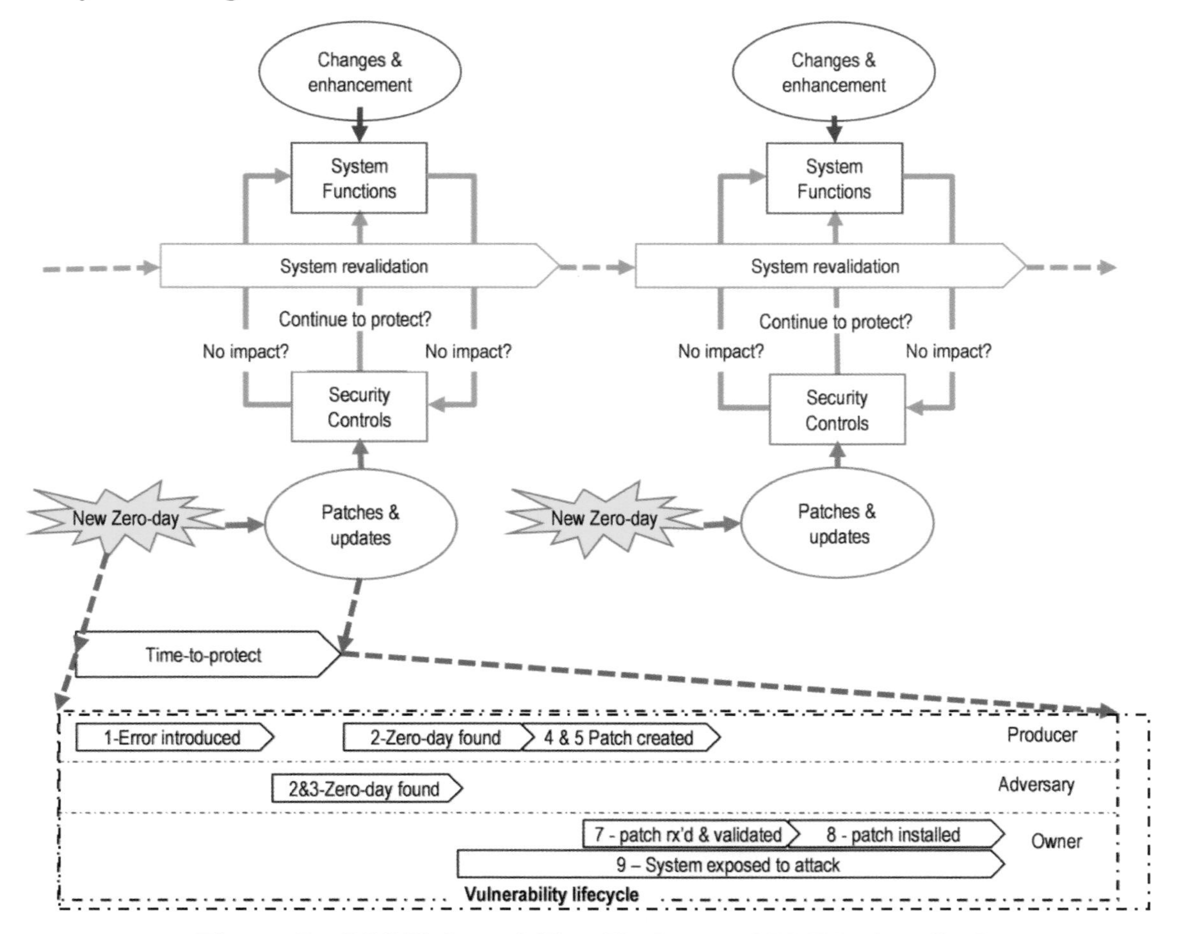


**Figure 7 ~ ICS Vulnerability, Update and Validation Cycle**

Vulnerability cycle includes:

1. point when software error results in vulnerability
2. point of discovery of new zero-day vulnerability

3. average time to exploitation in the field
4. time to develop a patch to address the vulnerability
5. time to communicate and act on vulnerability
6. time to stage patch and develop precautionary roll-back baseline
7. time to validate on test environment the reliability of the patch in the ICS and compatibility with Safety Functions
8. time to release into the production environment

The aim is to keep exposure time of the system, shown as 9 in Figure 7, as low as possible (hopefully before exploitation is initiated).

Patch Management in the Automation and Control System (IACS) environment; IEC 62443 (Parts 2-3) does specify methods and formats for patches in the development, notification, verification, and validation of security updates for OT systems (IEC n,d,).

### *2.6.2 CISA ICS Advisories and Alerts*

Awareness of emerging OT product vulnerabilities, and timely patching of these, is a key aspect of cybersecurity risk reduction. This requires a balance approach to ensure that system reliability is not compromised. US CISA provides timely advisories for ICS product vulnerabilities. Monitoring of these (https://www.cisa.gov/uscert/ics/advisories) and acting on advice should be a basis of maintain cybersecurity protection of ICS.

### *2.6.3 Physical Architecture and Network Segmentation*

Defence-in-depth, as provided by the long-accepted Purdue Enterprise Reference Architecture (Purdue 1989), is an example of a security architecture in critical industrial control system protected in "zones" and "conduits" (IEC n.d.). An example of this architecture is shown back in Figure 4 of Sub-section 2.4.

Network Segmentation has challenges in achieving separation of functions, let alone independence of these:

- We still need to communicate between IT and OT functions
  - OT functions need access to functionality in IT systems
  - IT systems need to monitor and control OT
- Network segmentation and functional separation can be bypassed by use of portable media (e.g. USB) or external maintenance links
- Firewalls separating network zones can be compromised with concerted efforts; if monitoring is only required across the gap the use of one-way firewalls or data-diodes may help in enforcing isolation.
- Security products are available that scan air-gapped systems without installing software, but these need to be set to not automatically delete files which could be critical to OT.

System Safety Standards, (IEC 2010), require that the boundary of the safety-related system be established and maintained. OT Cyber Security Standards (IEC n.d.) fulfil this by segmentation of safety related function into their own protected zones based on risk assessment, e.g. IEC 62443 Part 3.2.

Separation of functions across these safety boundaries must be maintained throughout the system lifecycle to prevent interference with process critical and safety-related functions (Hunter 2006).

### 2.7 Connectivity

Convergence of connectivity such as TCP/IP has increased the attack surface and vulnerabilities; however there have been previous attacks on systems with traditional OT architecture with safety and environmental impact, e.g. that on the Maroochy Shire Sewerage System (Smith 2001).

So, we know ICS systems cybersecurity's needs and practices; but what about protecting the safety-related elements?

## 3 Safe and Secure?

### 3.1 System Hierarchy

Safety Functions are rarely standalone, and are usually part of a larger system. In IACS, and their aligned systems, Safety Functions are considered, in the system hierarchy, as part of the control system essential functions as shown in Figure 8.

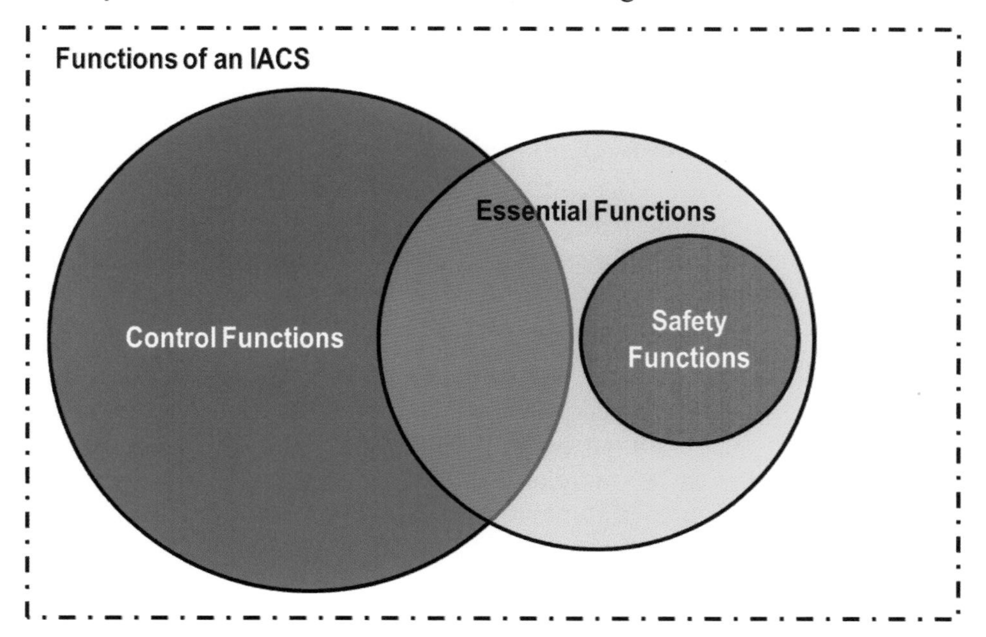


**Figure 8 ~ Hierarchy of IACS Functions (IEC 62443)**

This hierarchy helps to prioritise system dependability and protection from sources of unintended operation.

The following concepts, techniques and standards support ICS safety and security.

### 3.2 Boundaries are important after all!

As previously noted, safety standards usually call for the definition of the boundary of the safety-related elements and maintenance of and effective separation across this boundary (Hunter 2006). Setting security and safety boundaries has an impact on independence and interdependence of critical functions.

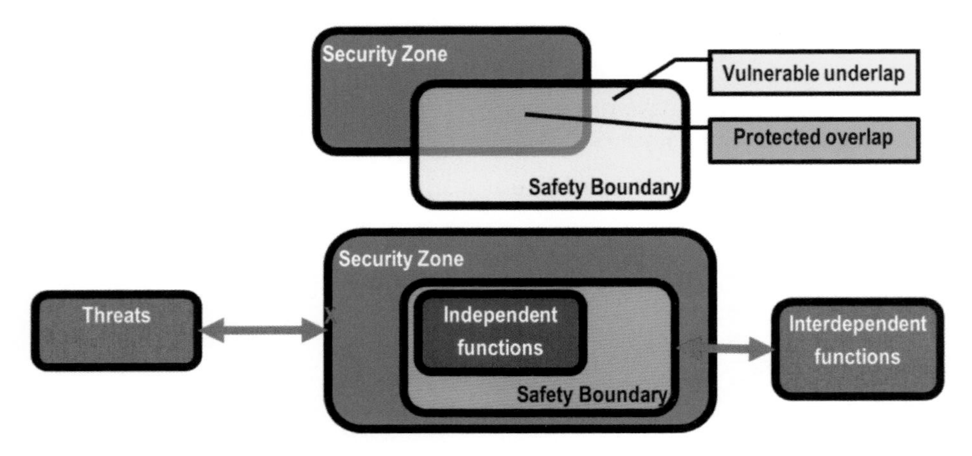


**Figure 9 ~ Safety/Security Boundary Considerations**

The placement of security perimeters and safety boundaries has a significant impact on the protection and reliability of safety functions:

- Overlap allows security countermeasures to work effectively and reduces the risk of incompatibility with safety functions;
- Underlap increases the risk of back-door attack vectors being exploited; and
- Reliability of safety functions may be affected by cybersecurity countermeasures inadvertently blocking interdependent functions outside the security perimeter.

## 3.3 Control Systems Kill Chain

Safety practitioners must face the reality that even with the best cybersecurity endeavours the possibility that Advanced Persistent Threat, "APT" will someday succeed in compromising their system. What is left is then reliance on effective safety and security response to this attack. To understand the best response to an attack, you need to know how attacks are staged.

The methods of cyber-attack have become advanced in application, and understanding the kill chain has helped to adapt mitigation to limit not only the likelihood of vulnerability exploitation but also its impact (Assante and Lee 2015). Figure 10 shows the relationship between the NIST Cybersecurity Framework (CSF) (NIST 2018a), Lockheed Martin's generic kill chain, and Safety System incident response/resiliency actions (IEC 2019a), (ISA 2017).

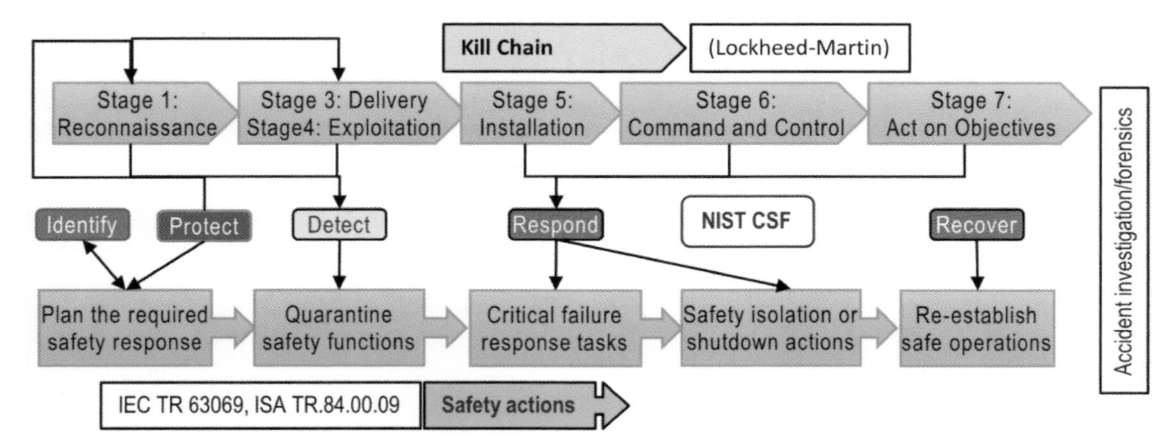


**Figure 10 ~ Interaction of Cybersecurity Kill Chain with NIST CSF and Safety**

Due to evolving nature of APT and despite best effects in cybersecurity, there is always a residual risk that safety-related components of a system will be compromised by a cyber-attack. Safety functions need to be resilient to this probability and ensure these do not lead to dangerous failures. Safety responses to a recognised attack include:

1. Plan the required safety response
2. Quarantine safety functions
3. Critical failure response tasks
4. Isolation or shutdown actions
5. Re-establish safe operations

Appendix A, Section A.1, provides a generic Framework Profile to address Ransomware risk of Safety Systems in OT.

## 3.4 Safety and Security Co-engineering Activities

Establishing safe and secure systems requires co-engineering to be undertaken across the system lifecycle (Paul et al. 2016).

IEC has published a Technical Report (IEC 2019a) to assist in protecting safety-related systems from dangerous cyber-attacks in applying safety (IEC 2010) and security (IEC n.d.) standards. It promotes the following guiding principles:

1. protection of safety implementations – to paraphrase, if it isn't secure, it isn't safe
2. protection of security implementations – to paraphrase, safety shouldn't increase security risk
3. compatibility of implementations – to paraphrase, security countermeasures shouldn't be unsafe

A fourth principle is being considered for a new edition with normative clauses:

- Guiding Principle 4: compatibility related to the higher-level system objectives
  - This could satisfy operational objectives, such as availability; and
  - Preclude down-time due to inadvertent shutdowns and fail-safe actions triggered by cyber-attacks.

Currently IEC 61508 is in development to be Edition 3. This includes more detail on handling cybersecurity risk to functional safety. This adds to clauses concerning Hazard and Risk Analysis in Edition 2 (IEC 2010).

ISA also has published a guide (ISA 2017) for the process sector to "address and provide guidance on the safety lifecycle and the cybersecurity lifecycle as they relate to the security of Safety Controls". It covers safety and security activities across the lifecycle from assessment, design, installation, operation, maintenance, modification, and decommissioning.

## 3.5 Unintended Countermeasure Consequences

Not all threats are external or adversarial. It is possible for inappropriate security countermeasures to impair the very functions they are protecting. This control conflict conundrum, as illustrated in Figure 11, has its counterparts in physical security such as fail-safe and fail-secure conflict of emergency exit doors (Hunter 2009) or physical security of aircraft cockpits (BEA 2016).

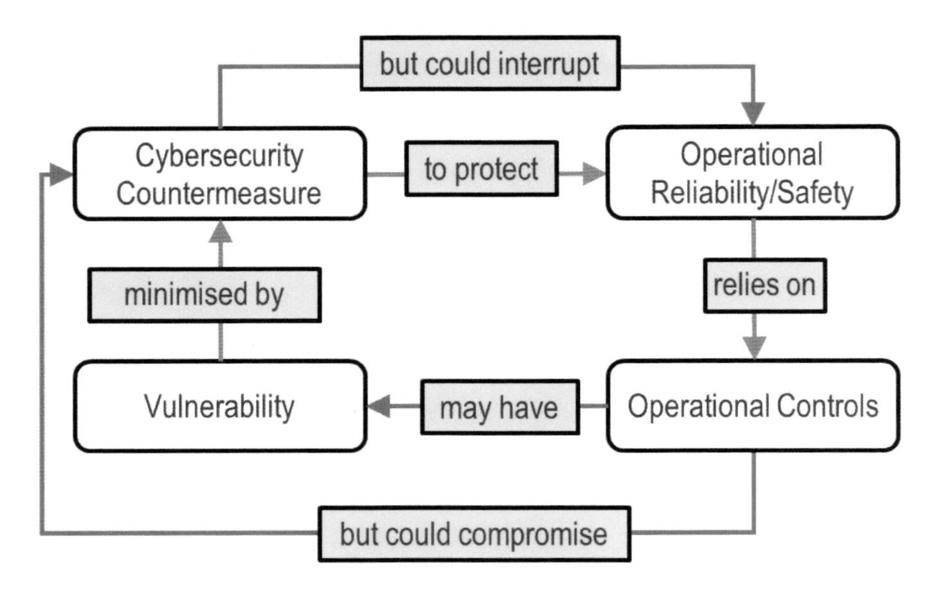


**Figure 11 ~ Control Conflict Conundrum**

The following safety and security standards do highlight this issue:

- NIST Cybersecurity Publications (NIST 2015), (NIST 2018b) classes these as "accidental or non-adversarial threats"
- Safety of Machinery IEC TR 63074 (IEC 2019b)
    - "*…shall not adversely affect safety integrity (e.g. increase in response time, etc.)*"
    - "*Any security countermeasure shall not adversely affect safety integrity (e.g. increase in response time, etc.)*"
- IEC TR 63069 (IEC 2019a) has a guiding principle: "*compatibility of implementations - Security implementations and safety implementations should not have adverse contradictions*"
- The ISA technical report on Cybersecurity Related to the Functional Safety Lifecycle (ISA 2017) classes these as "*security compromises* [which may] *occur during normal maintenance or other field activities where the cybersecurity compromise is unintentional or accidental*"

Appendix A, Section A.1, proposes common conflicts between cybersecurity and safety.

# 4 Mind the gap! - Maintaining Safety Independence

## 4.1 Risks to Functional Separation

Much reliance is placed on air-gaps, but even with no network connection between safety and non-safety related systems these can be bypassed. Loss of functional separation is almost entropic. The risk to this separation includes:

- Bypasses by human action including use of removable media, e.g. USB, that allow malware to infiltrate across the gap by breaking network segmentation and functional separation controls
- Unapproved or forgotten network connections including remote maintenance ports
- Untested changes to network configuration including:
    - message filtering;

- o port assignments;
- o access rules; and
- o authentication.

## 4.2 Establishing Effective Functional Safety Separation

Effective separation of safety systems is not as easy as it sounds. Strictly, the integrity of the separation mechanisms should have the same integrity as the safety system itself (Hunter 2006). This would mean:

- Network segmentation would require firewalls with a "SIL", i.e. Safety Integrity Level (IEC 2010)
- Human introduced bypasses, such as USB drives across air-gaps, would entail Human Reliability Analysis
- Hard-wired links across safety boundary may provide a level of isolation but require substantiation of meeting requirements for independence of safety, e.g. IEC 61508 Part 3, 7.4.2.9 (IEC 2010).

Pragmatically, separation will not be perfect or lasting; the fall-back position is to safely handle cybersecurity intrusions by a safety-driven incident response.

From a safety perspective, there are key points in the lifecycle to establish and maintain separation from non-safety parts of the system, as proposed in Table 1.

**Table 1 ~ Lifecycle Consideration of Safety Boundaries and Functional Separation**

| IEC 61508 Safety Lifecycle | Functional Safety Separation Activities | Related Cybersecurity Activities |
|---|---|---|
| Phase 4. Overall Safety Require-ments | Determine safety boundaries | Determine security perimeters and zones |
| Phase 5. Overall Safety Require-ment Allocation | Determine separation require-ments | Cybersecurity architecture segmen-tation requirements |
| Phase 9. System Safety Require-ments Specifica-tion | Specify trans-boundary infor-mation allowed and prohibited | Establish network firewall rules and air-gap needs |
| Phase 10. Safety-related Systems Realisation | Establishment of separation measures | Verify security requirements sup-port safety requirements. |
| Phase 13. Overall Safety Validation | Proof of separation of non-safety systems or influences | Validate effectiveness of security architecture and separation of safe-ty zones. |
| Phase 14. Overall Operation, Maintenance and Repair | Monitoring for compromised separation | Conduct safe ongoing dependency, visibility, and penetration testing |

| IEC 61508 Safety Lifecycle | Functional Safety Separation Activities | Related Cybersecurity Activities |
|---|---|---|
| Phase 15. Overall Modification and Retrofit | Re-evaluating safety boundaries and separation | Validate system modifications have not reduced the effectiveness of security architecture and separation of safety zones. Prevent unauthenticated SIS configuration changes. |

# 5 Summing up

Safety related systems are a growing target of cyber-attacks, including ransomware. Designers, installers, operators, and maintainers of these system should:

- Understand that safety functions in control systems are subject to increasing cyber threats;
- Ensure safety-related systems are effectively protected by well-maintained cybersecurity measures that are compatible with the system's safety functions;
- Safety and security must be coordinated and cooperative:
  - Safety practitioners, try talking with your cybersecurity counterparts
  - Cybersecurity practitioners, try talking with your system safety counterparts
  - You will be amazed how useful these conversations are…
- Segregation between OT and IT is not assured. Air-gapping is not certain (see RSA 2FA lesson (Greenberg 2021)). Functional separation must be proven and actively maintained for ongoing integrity of safety;
- The challenges of Software Defined Perimeters, such as used in IIoT and Factory 4.0, may increase cyber-attack surface for OT and safety-related systems. They may also decrease true reliability and provable safety integrity.

### Disclaimers

Opinions expressed in this paper are based purely on the public references listed. They do not infer culpability or carelessness on any party.

### Acknowledgments

The author acknowledges the dedication of safety and security professionals who protect us from dangerous outcome of failures in technology in our lives. The author appreciates the help of Gurinder Pal Singh, Ian H. Gibson, and anonymous peer-reviewers, who have offered advice on this article.

**References**

Assante, M. J., & Lee, R. M. (2015). *The Industrial Control System Cyber Kill Chain.* SANS White Paper, October 2015. Retrieved from SANS Institute: https://sansorg.egnyte.com/dl/HHa9fCekmc Accessed 19th January 2022

Bakuei, M., Flores, R., Remorin, L., & Yarochkin, F. (2021). *2020 Report on Threats Affecting ICS Endpoints.* Retrieved from Trend Micro Research: https://documents.trendmicro.com/assets/white_papers/wp-2020-report-on-threats-affecting-critical-industrial-endpoints.pdf Accessed 18th January 2022

BEA. (2016). *Final Report: Accident on 24 March 2015 at Prads-Haute-Bléone (Alpes-de-Haute-Provence, France) to the Airbus A320-211 registered D-AIPX operated by Germanwings.* Bureau d'Enquêtes et d'Analyses pour la sécurité de l'aviation civile, France. Final Report BEA2015-0125, March 2016. Retrieved from: https://www.bea.aero/uploads/tx_elydbrapports/BEA2015-0125.en-LR.pdf Accessed 20th January 2022

Biden, J. R. (2021). *National Security Memorandum on Improving Cybersecurity for Critical Infrastructure Control Systems.* July 2021. Retrieved from The White House: https://www.whitehouse.gov/briefing-room/statements-releases/2021/07/28/national-security-memorandum-on-improving-cybersecurity-for-critical-infrastructure-control-systems/ Accessed 30th July 2021

Carmakal C. (2021). Prepared Statement to the US House Committee on Homeland Security, June 9, 2021. Retrieved from https://homeland.house.gov/imo/media/doc/2021-06-09-HRG-Testimony%20Carmakal.pdf Accessed 18th January 2022

CISA. (2018). *ICS Focused Malware.* US Cybersecurity and Infrastructure Security Agency, ICS Alert (ICS-ALERT-14-176-02A), updated August 2018. Retrieved from: https://www.cisa.gov/uscert/ics/alerts/ICS-ALERT-14-176-02A Accessed 19th January 2022

CISA. (2019). *HatMan—Safety System Targeted Malware.* US Cybersecurity and Infrastructure Security Agency, Malware Analysis Report MAR-17-352-01, updated February 2019. Retrieved from: https://www.cisa.gov/uscert/sites/default/files/documents/MAR-17-352-01%20HatMan%20-%20Safety%20System%20Targeted%20Malware%20%28Update%20B%29.pdf Accessed 19th January 2022

CISA. (2020). *Ransomware Impacting Pipeline Operations*. US Cybersecurity and Infrastructure Security Agency, Alert (AA20-049A), updated October 2021. Retrieved from https://www.cisa.gov/uscert/ncas/alerts/aa20-049a Accessed 18th January 2022

CISA. (2021a). *Shamoon/DistTrack Malware.* US Cybersecurity and Infrastructure Security Agency, ICS Joint Security Awareness Report (JSAR-12-241-01B), updated July 2021. Retrieved from: https://www.cisa.gov/uscert/ics/jsar/JSAR-12-241-01B Accessed 19th January 2022

CISA. (2021b). *Ongoing Sophisticated Malware Campaign Compromising ICS.* US Cybersecurity and Infrastructure Security Agency, ICS Alert (ICS-ALERT-14-281-01E), updated July 2021. Retrieved from: https://www.cisa.gov/uscert/ics/alerts/ICS-ALERT-14-281-01B Accessed 19th January 2022

CISA. (2021c). *Cyber-Attack Against Ukrainian Critical Infrastructure*. US Cybersecurity and Infrastructure Security Agency, ICS Alert (IR-ALERT-H-16-056-01), updated July 2021. Retrieved from: https://www.cisa.gov/uscert/ics/alerts/IR-ALERT-H-16-056-01 Accessed 19th January 2022

CISA. (2021d). *CrashOverride Malware.* US Cybersecurity and Infrastructure Security Agency, Alert (TA17-163A), updated July 2021. Retrieved from: https://www.cisa.gov/uscert/ncas/alerts/TA17-163A Accessed 19th January 2022

CISA & FBI. (2021a). *Joint Cybersecurity Advisory: Chinese Gas Pipeline Intrusion Campaign, 2011 to 2013.* US Federal Bureau of Investigation and Cybersecurity and Infrastructure Security Agency, Alert (AA21-201A), updated July 2021. Retrieved from https://www.cisa.gov/uscert/ncas/alerts/aa21-201a Accessed 18th January 2022

CISA & FBI. (2021b*). DarkSide Ransomware: Best Practices for Preventing Business Disruption from Ransomware Attacks.* US Federal Bureau of Investigation and Cybersecurity and Infrastructure Security Agency, Alert (AA21-131A), updated July 2021, Retrieved from: https://us-cert.cisa.gov/ncas/alerts/aa21-131a Accessed 30th July 2021

GAO. (2018). *Critical Infrastructure Protection: Actions Needed to Address Significant Weaknesses in TSA's Pipeline Security Program Management.* United States Government Accountability Office. GAO-19-48, December 2018. Retrieved from https://www.gao.gov/assets/gao-19-48.pdf Accessed 18th January 2022

Goodin, D. (2010). *McAfee false positive bricks enterprise PCs worldwide.* April 2010. Retrieved from: https://www.theregister.com/2010/04/21/mcafee_false_positive/ Accessed 30th July 2021

Greenberg, A. (2021). *The Full Story of the Stunning RSA Hack Can Finally Be Told.* May 2021. Retrieved from WIRED: https://www.wired.com/story/the-full-story-of-the-stunning-rsa-hack-can-finally-be-told/ Accessed 30th July 2021

Hoffman, M & Winston, T. (2021). *Recommendations Following the Colonial Pipeline Cyber Attack.* Retrieved from Dragos website: https://www.dragos.com/blog/industry-news/recommendations-following-the-colonial-pipeline-cyber-attack/ Accessed18th January 2022

Hunter, B. (2006). *Assuring Separation of Safety and Non-safety Related Systems.* 11th Australian Workshop on Safety Related Programmable Systems (SCS'06), Melbourne: Conferences in Research and Practice in Information Technology, Vol. 69. Retrieved from: https://dl.acm.org/doi/pdf/10.5555/1274236.1274243 Accessed 19th January 2022

Hunter, B. (2009). *Integrating safety and security into the system lifecycle.* In Improving Systems and Software Engineering Conference (ISSEC), Canberra, Australia, p. 147. August 2009

Hunter, B. (2013). *Verifying Security-Control Requirements and Validating their Effectiveness.* INCOSE Insight, Volume 16 Issue 2, pp 45-48. June 2015

IEC (n.d). *Industrial communication networks - Network and system security*, IEC 62443, all parts separately dated. International Electrotechnical Commission, Geneva

IEC. (2010). *Functional safety of electrical/electronic/programmable electronic safety-related systems.* IEC 61508. International Electrotechnical Commission, Geneva.

IEC. (2019a). *Technical Report: Industrial-process measurement, control and automation – Framework for functional safety and security.* IEC TR 63069:2019, International Electrotechnical Commission, Geneva

IEC. (2019b). *Technical Report: Safety of machinery - Security aspects related to functional safety of safety-related control systems.* IEC TR 63074:2019, International Electrotechnical Commission, Geneva

ISA. (2017). Cybersecurity Related to the Functional Safety Lifecycle. ISA-TR84.00.09-2017, International Society for Automation, Research Triangle

ISACA. (2016). *The Merging of Cybersecurity and Operational Technology.* Information Systems Audit and Control Association (ISACA) White Paper.

ISACA. (2021). *ISACA Survey: IT Security and Risk Experts Share Ransomware Insights in the Aftermath of the Colonial Pipeline Attack.* May 24, 2021. Retrieved from ISACA: https://www.isaca.org/why-isaca/about-us/newsroom/press-releases/2021/it-security-and-risk-experts-share-ransomware-insights-in-the-colonial-pipeline-attack Accessed 30th July 2021

ISO/IEC. (2020). *Technical Report: Internet of things (IoT) — Industrial IoT.* ISO/IEC TR 30166:2020, International Organization for Standardization and International Electrotechnical Commission, Geneva

Kelso, M. (2020). *Pipelines Continue to Catch Fire and Explode*. Retrieved from: https://www.fractracker.org/2020/02/pipelines-continue-to-catch-fire-and-explode/ Accessed 30th July 2021

Moore, S. (2021). *Gartner Predicts By 2025 Cyber Attackers Will Have Weaponized Operational Technology Environments to Successfully Harm or Kill Humans.* Retrieved from: https://www.gartner.com/en/newsroom/press-releases/2021-07-21-gartner-predicts-by-2025-cyber-attackers-will-have-we Accessed 30th July 2021

NIST. (2015). Guide to Industrial Control Systems (ICS) Security. SP 800-82 Rev. 2, May 2015. Retrieved from National Institute of Standards and Technology: https://nvlpubs.nist.gov/nistpubs/SpecialPublications/NIST.SP.800-82r2.pdf Accessed 19th January 2022

NIST. (2018a). *Framework for Improving Critical Infrastructure Cybersecurity.* April 2018. Retrieved from National Institute of Standards and Technology: https://nvlpubs.nist.gov/nistpubs/cswp/nist.cswp.04162018.pdf Accessed 19th January 2022

NIST. (2018b). *Systems Security Engineering: Considerations for a Multidisciplinary Approach in the Engineering of Trustworthy Secure Systems.* SP 800-160 Vol 1, updated March 2018. Retrieved from National Institute of Standards and Technology: https://nvlpubs.nist.gov/nistpubs/SpecialPublications/NIST.SP.800-160v1.pdf Accessed 19th January 2022

NIST. (2021). *Cybersecurity Framework Profile for Ransomware Risk Management.* NISTIR 8374 (Draft, September 2021). Retrieved from National Institute of Standards and Technology: https://nvlpubs.nist.gov/nistpubs/ir/2021/NIST.IR.8374-draft.pdf Accessed 19th January 2022

Palmer, D. (2021). *Ransomware gangs are taking aim at 'soft target' industrial control systems.* July 2021. Retrieved from ZDnet Security website: https://www.zdnet.com/article/ransomware-gangs-are-taking-aim-at-soft-target-industrial-control-systems/ Accessed 30th July 2021

Parfomak, P.W. (2012). *Pipeline Cybersecurity: Federal Policy*. Congressional Research Service Report for Congress, R42660, August 16, 2012. Retrieved from https://sgp.fas.org/crs/homesec/R42660.pdf Accessed 18th January 2022

Paul, S., Rioux, L., Gailliard, G., & Wiander, T. (2016). *Recommendations for Security and Safety Co-engineering.* The Information Technology for European Advancement (ITEA 2) project "MERgE"

Purdue Research Foundation (T. J. Williams). (1989). *A Reference Model For Computer Integrated Manufacturing (CIM).* December 1989. Instrument Society of America, Research Triangle

Smith, T. (2001). *Hacker jailed for revenge sewage attacks*. October 2001. Retrieved from: https://www.theregister.com/2001/10/31/hacker_jailed_for_revenge_sewage/ Accessed 19th January 2022. For more detail of the referenced court case, see also https://www.queenslandjudgments.com.au/caselaw/qca/2002/164 Accessed 30th July 2021

Timpson, D., & Moradian, E. (2018). *A Methodology to Enhance Industrial Control System Security.* Proceedings of the 22nd International Conference on Knowledge-Based and Intelligent Information & Engineering Systems, Belgrade, Serbia (pp. 2117-2126). Elselvier & sciencedirect.com

TSA. (2021a). *Pipeline Security Guidelines*. Retrieved from US Transport Security Administration: https://www.tsa.gov/sites/default/files/pipeline_security_guidelines.pdf Accessed 30th July 2021

TSA. (2021b). *Enhancing Pipeline Cybersecurity.* Security Directive Pipeline-2021-01. May 2021. Retrieved from: https://www.powermag.com/wp-content/uploads/2021/05/sd-pipeline-202-1-01-tsa.pdf Accessed 18th January 2022

# Appendix A. Supplemental Material

## A.1 Countermeasure Safety Issues

Inappropriately configured or utilised cybersecurity countermeasure can have negative impacts on safety as summarised by Table 2.

**Table 2 ~ Cybersecurity and Safety Conflict Issues**

| **Countermeasure/ Activity** | **Risk to Safety Function** | **Possible Mitigations** |
|---|---|---|
| Penetration testing | Could disrupt safety system or cause uncommanded dangerous operation | Have safe and proven penetration testing tools – isolate dangerous operation |
| Patching incompatibility | Could disrupt safety system or cause uncommanded dangerous operation | Verify patch in pre-production platform |
| AV false positive | Could stop safety functions | Verify anti-virus update in pre-production platform |
| PKI certificates expiry | Could stop safety functions | Ensure safety functions not compromised by authentication failures |
| Firewall policy changes | Could stop safety communications or dependencies | Validate and control firewall policy especially in safety conduits |
| Password expiry policy | Enforced timeout on passwords may stop operator from enacting safety-related commands | Ensure effective access control management that maintains safety controls |
| Intrusion Protection System (IPS), Endpoint Detection and Response (EDR & XDR) | Could stop safety functions if these are visibly confused with cyber-attack | Isolate IPS, Endpoint Detection and Response (EDR) and Extended Detection and Response (XDR) systems from critical safety zones |
| Networked or portable malware scanning and removal tools | False positives could remove OT critical files and applications | Validate tool in OT sandbox or test environment. Set to delete files manually. |
| Online control system internet presence discovery tools, e.g. Shodan, Google dorks | Could expose safety-system addressing to threat actors. | Use offline tools instead, e.g. NMAP, ZENMAP |

## A.2 Generic Framework Profile for Safety Systems

The NIST Cybersecurity Framework (NIST 2018a) provides a complete security cycle perspective for aspects of identification, protection, detection, response, and recovery to maintain the cybersecurity of systems. NIST has published a draft CSF profile for ransomware of critical infrastructure (NIST 2021) which provides a tailoring to the profile application to Ransomware (along with other Malware) and System Safety mitigation in Operational Technology (OT). This can be useful to assess countermeasures applied to these systems and threats they address. This framework allows specific profiles to be tailored to a system and its cybersecurity risks. Table 3 proposes safety considerations for CSF cycles and categories.

**Table 3 ~ Framework Profile Considerations for Safety**

| NIST CSF Cycle | Category | Functional Safety Considerations |
|---|---|---|
| Identify (ID) | Asset Management (ID.AM) | Safety assets and configuration identified. |
| | Business Environment (ID.BE) | Organizational safety roles and responsibilities assigned |
| | Governance (ID.GV) | Safety regulatory requirements established. |
| | Risk Assessment (ID.RA) | Safety included in hierarchy of risk assessment. |
| | Supply Chain Risk Management (ID.SC) | Response and recovery plans are tested to include safety responses. |
| Protect (PR) | Identity Management, Authentication and Access Control (PR.AC) | Network segmentation and air-gapping of safety related systems established and maintained |
| | Awareness and Training (PR.AT) | Awareness and training include interaction between safety and security responsibilities. |
| | Data Security (PR.DS) | Penetration and visibility testing is conducted with safety in mind |
| | Information Protection Processes and Procedures (PR.IP) | Response and recovery plans regularly tested with safety and cybersecurity actions<br>Vulnerability management and patching includes impact and integrity of safety systems |
| | Maintenance (PR.MA) | Changes to safety system are approved, logged, and performed in a manner that prevents unauthorized access |
| | Protective Technology (PR.PT) | Fail-safe and fail-secure mechanisms are established to deal with dangerous cyber events. |

| NIST CSF Cycle | Category | Functional Safety Considerations |
|---|---|---|
| Detect (DE) | Anomalies and Events (DE.AE) | System safety is included in determination of event impact. |
| | Security Continuous Monitoring (DE.CM) | Isolated safety zones are monitored for evidence of intrusion. |
| | Detection Processes (DE.DP) | Cybersecurity intrusion detection and protection systems do not compromise safety functions, e.g. false positives |
| Respond (RS) | Response Planning (RS.RP) | Joint safety and security response plans are executed to protect safety during attack. |
| | Communications (RS.CO) | Information is shared between safety and security personnel and coordination is practiced to minimise dangerous failures. |
| | Analysis (RS.AN) | Safety and security personnel cooperate on forensics and accident analysis that results from event. |
| | Mitigation (RS.MI) | Safety and security personnel coordinate mitigation of the event including precautionary shutdowns |
| | Improvements (RS.IM) | Safety and security personnel cooperate on lessons learned from the event. |
| Recover (RC) | Recovery Planning (RC.RP | Restoration plans are coordinated between safety and security. |
| | Improvements (RC.IM) | Lessons learned are included into response plans and processes. Gaps in safety and security protections are acted on. |
| | Communications (RC.CO) | Restoration plans are coordinated between safety and security including return to safe-state operations. |

This collation page left blank intentionally.

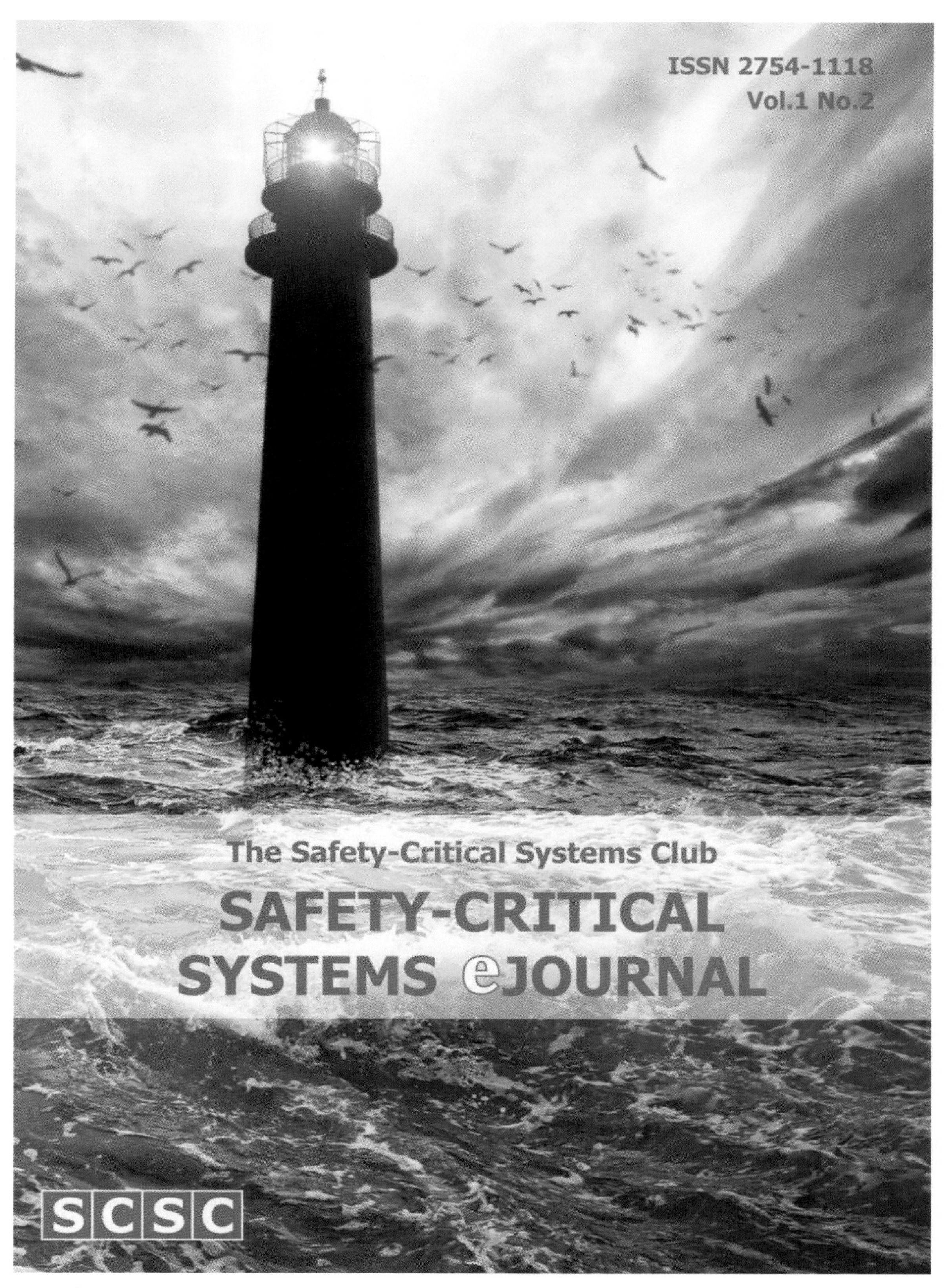


Issue 2 Cover

This collation page left blank intentionally.

ISSN 2754-1118 (Online) — ISSN 2753-6599 (Print)

# Editorial to the Second Issue

This second issue of the Safety-Critical Systems eJournal, which is published by the Safety Critical Systems Club, clearly fulfils one of the purposes set out in the original brief for the journal. That is to provide a home for articles and papers that are too long for the Newsletter and Symposium. It illustrates two ways of presenting larger amounts of material: one paper is the first part of three, the other is a long paper, which may be built upon in later publications. They are:

- Derek Fowler (UK), in "*IEC 61508 Viewpoint on System Safety in the Transport Sector: Part 1 – An Overview of IEC 61508*", considers how the safety assessment of new (or significantly modified) systems deployed in specific transport applications would look if it followed, as far as practicable, the safety-specification phases of the lifecycle set out in Part 1 of IEC61508.

- Nicholas Hales (UK), in "*The Layered Enterprise Data Safety Model (LEDSM): A Framework for Assuring Safety-critical Communications*", considers the lack of advanced planning in safety-critical networks, which has made the outcomes of a number of incidents more severe than they needed to have been. He proposes a way to develop safer networks of communication of any type, verbal, telephone, internet, etc., or a mix of types, so that the risks of failures to communicate are considerably reduced. Since this paper was received, we have seen other incidents wherein poor communications has been an aggravating factor, such as a shooting at the Robb Elementary School in Texas and crowd management chaos at the Stade de France, Paris.

The editorial to the first issue of the journal said, "*You may find some of this material controversial, or you may think that it does not go far enough. Subsequent issues of this journal will have provision for readers' letters to the Editor responding to individual papers.*" Such a letter has been received from Bev Littlewood and Lorenzo Strigini, "*A Critical Response to a Recent Paper by Daniels and Tudor*". Also received on that topic is a short paper:

- Peter Ladkin (Germany), in "*Evaluating Software Execution as a Bernoulli Process*", sets out to show that examples and considerations cited by Daniels and Tudor in their paper do not follow the constraints necessary for modelling any process as a Bernoulli Process, hence challenging their conclusions.

My thanks go to the authors for contributing their papers, and also to the peer-reviewers (at least three per paper) for suggesting improvements. Apologies also to those reviewers who made some recommendations that were not taken up.

It is our intention to make Issue 1 of Volume 3 a special issue addressing certain aspects of autonomous vehicle safety. More details of this in the next issue…

John Spriggs, SCSC Journal Editor

July 2022

ISSN 2754-1118 (Online) — ISSN 2753-6599 (Print)

# Letter to the Editor

**From: Bev Littlewood and Lorenzo Strigini**

**Subject: A Critical Response to a Recent Paper by Daniels and Tudor**

It is good to see the new Safety-Critical Systems Club eJournal's first issue contributing to debate about the use of probabilistic models to evaluate the dependability of software-based systems. Such evaluation is particularly important for safety-critical systems, especially those whose failures may have massive consequences. Society needs assurances that such systems are "good enough". It is therefore appropriate that methods for providing these assurances are subject to critical examination, as Daniels and Tudor do here (Daniels and Tudor 2022) — hereafter D&T for brevity.

D&T refer to moves within a civil aviation technical community towards giving more weight, in assessment, to records of good behaviour of the product, and perhaps less weight to records of precautions in development. They do not give enough detail for others to agree or not on whether these moves would be good or bad. We thus limit our discussion to their general claims against use of statical testing and operational evidence.

Unfortunately, whilst some of the D&T claims *are* valid, we think some are not. For example, they cite the requirement for some failures of systems in aircraft to be extremely improbable. They rightly point out that the required levels cannot be assured via statistical modelling — they cite our own old paper (Littlewood and Strigini 1993) and that of others (Butler and Finelli 1993) in support. However, they go on to claim that they have identified "*an alternative way forward*" to statistical reasoning "*that does provide evidence that software is safe for its intended use before it enters service*". Both some of their opposition to statistical argument, and this latter claim, seem wrong, as we argue below. Some of our reasoning is reprised from our later paper (Littlewood and Strigini 2011).

There are some levels of reliability requirement (what we called "ultra-high" in our 1993 title) that cannot be demonstrated in advance to be satisfied. We believe this will remain true. However, statistical and probabilistic reasoning helps to determine what one can reasonably believe, and thus to answer the question, among others, whether the level of risk associated with a system is socially tolerable.

The D&T objections to statistical testing mix matters of *model misuse*, of *model validity* and of *adequacy of evidence*.

Some of their arguments are about misuse of methods. In their example of a system whose behaviour depends on the date, assessment indeed requires a proper sample of the space of all dates on which it will be used. Not procuring this, as they hypothesise, is a textbook mistake. Although they do not cite instances of such misuse occurring in practice, if it were common, it would be indeed a concern; but not a reason for giving up the advantages of statistical evidence and statistical reasoning.

"Model validity" concerns the first stage in evaluation of reliability: building a probabilistic model that represents, with acceptable approximation, the real-world properties of a stochastic process of failures. One such model is the Bernoulli Process (BP) for demand-based systems, which D&T criticise in some detail as inadequate for many real-world situations. In the BP model a system responds to a series of "demands", and each response may be either a *failure* or a *success*. The model is based on two assumptions: the outcomes of executions on successive demands are statistically independent, and the probability of failure on a demand, $p$, is constant for the successive demands. These assumptions are, of course, not always appropriate. They are, though, in many practical situations. For example, in the long association we have had with scientists and engineers in the nuclear industry, there has been wide acceptance of the BP model for reactor protection systems. We believe this is reasonable, because the two underpinning assumptions seem plausible there: demands are infrequent, so it can be expected that the outcome of a demand will be independent of the outcome of the previous demand, that might have taken place a year earlier; and there are good arguments for assuming $p$ to remain approximately constant over some extent of time.

BP models are indeed not always appropriate, and for these situations there is now an extensive literature about more general models that address such issues, for example by relaxing the assumption of constant $p$. D&T do not address these, or other even simpler solutions for the problems that they cite as showing that BP models for software failures are "*fundamentally flawed*". For example, how to deal with software that has state has been publicly explained at least as far back as in our 1997 work for ESA, the European Space Agency (Strigini and Littlewood 1997).

"Adequacy of evidence" concerns the next, statistical stage of evaluation. For instance, in BP models, one uses collected evidence (successes and failures of a system, in operation or in statistical testing) to improve one's estimate of the parameter $p$, the (constant) probability of failure on demand. This will be unknown, and statistical inference about its value is needed to support probabilistic predictions about future failure behaviour of the system. With safety-critical systems, there is particular interest in the situation where there have been very many demands with no failures. This represents "best news" statistical evidence in support of a claim that a system is "good enough". See for example the Table from IEC 16508 Annex D to Part 7 which D&T reproduce[1]. D&T rightly give particular attention to this special case. They remind the reader of what many, we among them, have been pointing out: there are limits to how much evidence of failure-free execution can realistically be obtained, with consequent limits on the levels of reliability that can be assured. Thus when, for instance, a requirement of extreme improbability of certain failures is explained as "Extremely improbable failure conditions are those having a probability on the order of $1 \times 10^{-9}$ or less" (as in FAA AC-1309-A, and similar documents), collecting enough evidence is currently seen as infeasible.

On the other hand, not all safety-critical systems have ultra-high reliability requirements. IEC 61508 does consider low SIL levels, for instance. The software-based primary protection system of the Sizewell B reactor was only required to have a probability of failure on demand no worse than $10^{-3}$ (there were other protections in the wider systems, including a hard-wired secondary system). This goal was sufficiently modest that it was eventually demonstrated with high confidence via statistical testing.

So, what can be done about systems with unassurable ultra-high reliability requirements, such as those for airplanes?

We agree with D&T that the record of in-flight operation of some critical systems, over many years, is extraordinarily impressive. So it seems that ultra-high reliability *may* have been *achieved* here, as evidenced after massive operational use. But this is rather different from claiming high confidence in such ultra-high reliability *before* a system enters service.

[1] Note that this table from the IEC standard presents *2-sided* confidence intervals for $p$. In fact they should instead be *1-sided* confidence bounds: a user wishes to know how confident they can be, for given evidence, that $p$ is *smaller* than a certain bound. They have no interest in knowing how confident they can be that $p$ is larger than a bound. D&T have no comment on this.

Nevertheless D&T say: "We now have nearly 30 years of service experience that satisfying the objectives of RTCA/DO-178B/C has been sufficient to address the software considerations in aircraft certification."

And go on to say: "…this paper proposes an alternative way forward that does provide evidence that *software is safe for its intended use before it enters service*." (Our italics)

We found the authors' reasoning at this point rather vague and hard to follow. The argument that 30 years of experience prove something *is* a statistical argument, but stated in a rather hand-waving manner.

D&T may mean simply that regulators agree to accept application of RTCA/DO-178B/C in lieu of evidence of satisfying $10^{-9}$: a simple fact. Or they may mean that applying the standards assure that result, so that it can be taken as *evidence* of having *achieved* it. This is a bold claim. Can they actually prove it? To do so, they would need to show that the $10^{-9}$ objective has indeed been achieved, consistently over most systems (a hard claim to demonstrate for most of them); next, that the attainment is linked to applying RTCA/DO-178B/C prescriptions in such a way that we should consider RTCA/DO-178B/C compliance sufficient evidence. At this point they could claim a *probability* that the methods will produce satisfaction of the requirements in the next aircraft type developed. This probability, dependent on the numbers of such successes and of systems, will be a rough estimate or range, certainly, but would help to see how strong their claim is.

Evidence of good practice and diverse forms of verification is indeed a valid part of an argument for high reliability or safety. Various authors, ourselves included, have proposed ways for clarifying *how much* they can contribute to sound quantitative arguments for reliability or safety; see for example, Bishop, Bloomfield, et al, (2011), Strigini & Povyakalo (2013) and Littlewood, Salako, et al (2020). A rough estimate of probability of achieving the requirement through application of RTCA/DO-178B/C precautions would fit well in such reasoning. It would almost certainly still imply an excessive probability of catastrophic failures in a type's lifetime, yet this could be reduced with statistical evidence from testing and operation.

An especially negative effect of D&T's argument is that, while they oppose statistical reasoning for ultra-high requirements (because – paraphrasing – it will refuse to deliver the reassuring statements that one may wish to hear, albeit giving *useful* information about *how much* has actually been demonstrated), they then decry statistical methods much more generally. Yet for many systems with more modest reliability and safety requirements, quantitative assurance can indeed be effectively obtained with statistical testing and simple probability models, such as the Bernoulli Process (and its continuous time equivalent, the Poisson Process). Even safety-critical systems often fall into this category (e.g. see our earlier example of the Sizewell B nuclear reactor protection system). Thus D&T's broad opposition to reliability modelling for software runs the risk of encouraging system builders to eschew proper quantitative evaluation in favour of informal qualitative arguments: e.g. "you can trust this system to be safe enough, because we used accepted best practice in building it." Such hand-waving justification is not good enough for highly critical systems.

(Emeritus Prof) Bev Littlewood

(Prof) Lorenzo Strigini

Centre for Software Reliability

City, University of London

29 May 2022

**References**

Bishop, P. G., Bloomfield, R. E., Littlewood B., Povyakalo A. and Wright D. R. (2011). *Toward a Formalism for Conservative Claims about the Dependability of Software-Based Systems*. IEEE Trans Software Engineering, 37(5), pp. 708-717. doi: 10.1109/TSE.2010.67

Butler, R. W. and Finelli, G. B. (1993). *The infeasibility of quantifying the reliability of life-critical real-time software*. IEEE Trans Software Engineering 19(1): 3-12.

Daniels, D. and Tudor, N. (2022). *Software Reliability and the Misuse of Statistics*. Safety-Critical Systems eJournal 1(1).

Littlewood, B. and Strigini, L. (1993). *Validation of ultra-high dependability for software-based systems*. CACM 36(11): 69-80.

Littlewood, B. and Strigini, L. (2011). *'Validation of Ultra-High Dependability' — 20 years on*. SCSC Newsletter 20(3).

Littlewood, B., Salako, K., Strigini, L. and Zhao, X. (2020). *On Reliability Assessment When a Software-based System Is Replaced by a Thought-to-be-Better One*. Reliability Engineering & System Safety, 197 [106752]. doi: 10.1016/j.ress.2019.106752

Strigini, L. and Littlewood, B. (1997). *Guidelines for Statistical Testing* (PASCON/WO6-CCN2/TN12). ESA/ESTEC project PASCON. https://openaccess.city.ac.uk/id/eprint/254/2/StatsTesting_TN12-3.1distrib2.pdf

Strigini, L. and Povyakalo, A. (2013). *Software fault-freeness and reliability predictions*. Proc. SAFECOMP 2013. (pp. 106-117). Cham: Springer. ISBN 978-3-642-40792-5

ISSN 2754-1118 (Online) — ISSN 2753-6599 (Print)

# IEC 61508 Viewpoint on System Safety in the Transport Sector

## Part 1 – An Overview of IEC 61508

**Derek Fowler**

Independent Safety Engineering Consultant, Reading, UK

## Abstract

*IEC publication 61508 "Functional safety of electrical/ electronic/programmable electronic safety related systems" is probably the most widely accepted, international generic standard on functional safety. Although its roots can be traced to process industries, the intention behind the Standard has always been to provide a solid, comprehensive basis for adaptation to a wide range of industry sectors. Nevertheless, previous published research into safety engineering practices in the transport sector has shown that, in some areas, those practices have failed to recognise even some of the most basic principles of IEC 61508 (as set out in Parts 1 and 4 of the Standard) and, as a consequence, focussed far more on the reliability of safety-related systems, and not enough on their potential risk-reduction properties. This paper, which is to be published as a series of parts, starting with this overview document (Part 1), will explore how IEC 61508 could be applied directly to the transport sector, with substantially beneficial results. No attempt is made to compare those results with actual practices in the specific transport applications addressed in Parts 2 and 3 – that is left to readers with a specific interest in those applications.*

# 1 Introduction

### 1.1 Background

IEC publication 61508 "*Functional safety of electrical/ electronic/programmable electronic safety related systems*" (IEC 2010) is probably the most widely accepted, international generic standard on functional safety. Although its roots can be traced to process industries, the intention behind the Standard has always been to provide a solid, comprehensive basis for adaptation to a wide range of industry sectors.

Using some simple principles from IEC 61508 as a benchmark, Fowler (2015) considered safety standards that were representative of practices in two transport-industry sectors — rail and aviation / air traffic management (ATM) — and found that, in some cases, they were "*wholly inadequate*" in that they focussed far more on the reliability of safety-related systems and not enough on their potential risk-reduction properties.

## 1.2 Aim and Objectives

The aim of this paper is quite different from, but complementary to, that of Fowler (2015). It seeks only to answer the simple question as to how, in the opinion of the author, the safety assessment of new (or significantly modified) systems deployed in specific transport applications would look if it followed, as far as practicable, the safety-specification phases of the lifecycle set out in Part 1 of IEC 61508, "IEC 61508-1". Any comparison of such an approach with current practices is left to the reader!

It is intended that the paper be issued in three parts, as follows:

**Part 1**, this document, whose objectives are to capture the essential principles that underpin IEC 61508, and show how those principles are embedded in the requirements-specification phases of the lifecycle set out in IEC 61508-1;

**Part 2** will describe, through a worked example, how the principles and lifecycle processes from Part 1 of the paper could be applied to European Air Traffic Management, and what the results would look like; and

**Part 3** will describe, through a worked example, how the principles and lifecycle processes from Part 1 of the paper could be applied to European rail transport, and what those results would look like.

## 1.3 Scope

The overall scope of the paper is deliberately limited to the safety-requirements specification phases of the IEC 61508 lifecycle. This is because most of the key principles underpinning IEC 61508, i.e. those universal principles set out in Parts 1 and 4 of the Standard, which govern the determination of the required risk-reducing properties of safety-related systems, take effect during these earlier phases, whereas the subsequent realisation and operating phases are less specific to the Standard.

Two important notes in IEC 61508-1, Sub-section 1.2, also have an impact on the scope of the discussions herein:

- Note 2: "*although a person can form part of a safety-related system (see also IEC 61508-4), human factor requirements related to the design of ... safety-related systems are not considered in detail in the standard*";
- Note 4: "*although the overall safety lifecycle is primarily concerned with [electronic / programmable] safety-related systems, it could also provide a technical framework for considering any safety-related system irrespective of the technology of that system*".

## 1.4 Document Layout

After this introductory section, Section 2 of this document sets out the key concepts and principles upon which IEC 61508 is based.

Section 3 then outlines how these concepts and principles are applied in the safety-requirements-specification phases of the IEC 61508 lifecycle, using the simple example of a hypothetical pedestrian-crossing facility to illustrate the ideas behind them.

Semantics is vital to a clear understanding of IEC 61508; therefore, Appendix A hereto contains firstly a set of common safety definitions and, secondly, a set of terms that have a particular meaning in relation to IEC 61508. Reference to a definition is given below thus: '[A.n-r]', where 'r' provides a hyperlink to the relevant item.

# 2 IEC 61508 Key Principles

## 2.1 Functional Safety Concepts

Functional safety [A.2-5], unlike "health and safety at work", is *[that] part of the overall safety relating to the EUC and the EUC Control System that depends on the correct functioning of the safety-related systems and other risk-reduction measures*.

Any safety assessment that accords with the key principles of IEC 61508 must start with the concept of the equipment under control (EUC). The definition of EUC [A.2-1]: "*equipment, machinery, apparatus or plant used for manufacturing, process, transportation, medical or other activities*", gives an insight into how the idea of an EUC might be interpreted more broadly than its actual title might suggest; for example, the flow of traffic along a road (or roads), the movement of trains along railway track, and the flow of aircraft through a portion of airspace, might all be thought as being an "EUC", and such a broader interpretation is not prohibited by the above definition. Whatever form the EUC takes, its essential property is that it is inherently hazardous!

An EUC Control System is defined as "*that system which responds to input signals from the [EUC] and/or from an operator, and generates output signals causing the EUC to operate in the desired manner*" [A.2-3]. Examples from the transport sector are road-vehicle control systems (including the driver, if not fully automatic), railway-signalling systems and air-traffic control systems. It could be an integral part of the EUC or a separate system and could, if it made a significant contribution to the safety of the EUC, be considered to be a safety-related system [A.2-10] in its own right.

From a safety perspective, two crucial points need to be born in mind:

- firstly, it is the EUC, together with its Control System, which lies at the heart of the IEC 61508 safety assessment process, since it is the (unmitigated) risk caused by the very existence of the EUC, for which safety functions [A.2-7] [A.2-9] need to be provided in order for the EUC Risk [A.2-4] to be reduced to a Tolerable level [A.1-5]; and
- secondly, whether an overall functional system is safe or not depends not just on its own inherent properties but also on the parameters of the Environment [A.2-2] —including, for example, physical, operating, legal and maintenance environments — in which the system is deployed.

## 2.2 Risk Reduction Overview

The principle of risk reduction is illustrated first of all in Figure 1, which is a generalised risk model, adapted from Figure A.1 of Part 5 of IEC 61508, "IEC 61508-5".

In the definition of EUC Risk [A.2-4], Note 4 emphasises that the "*main purpose of determining the EUC Risk is to establish a reference point for the risk without taking into account* [the possible risk reduction afforded by] *any Safety-related Systems or any other risk-reduction measures*"[2].

[2] In *practice*, we will see that it is not always essential (or possible) to determine an absolute value of EUC Risk *provided*: all of the hazardous events associated with the EUC, and its operational environment, are identified, appropriate safety functions are specified, and the residual risk can be shown to be less that the relevant tolerable risk.

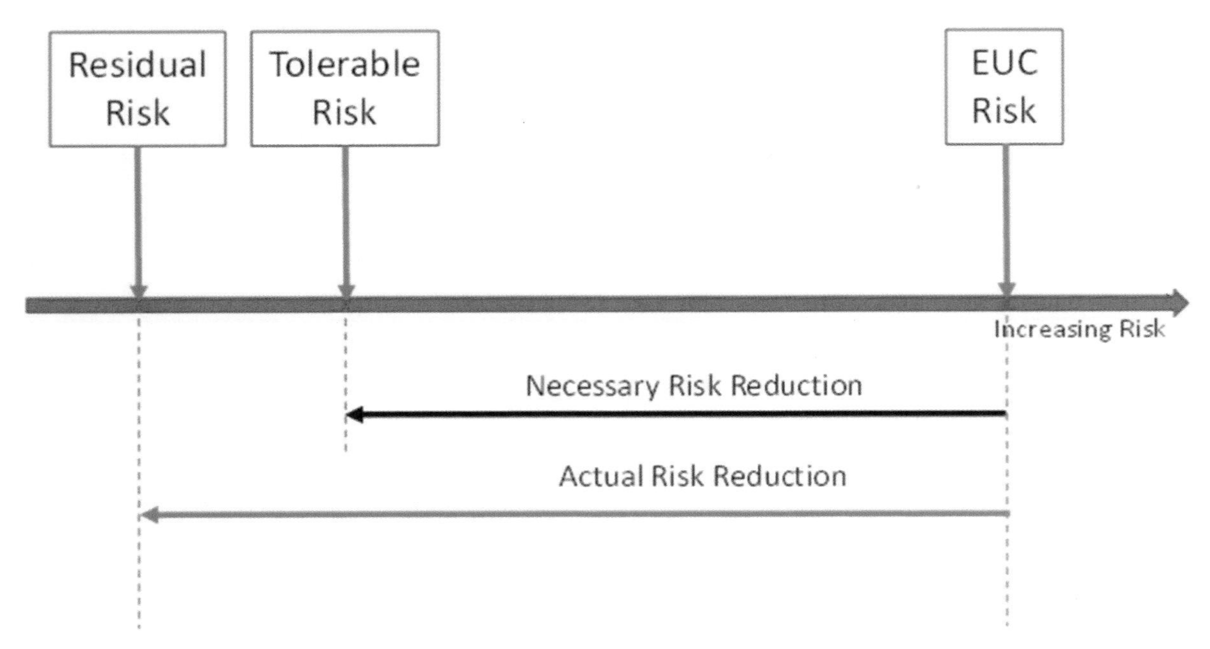


**Figure 1 ~ Basic Risk Model**

Necessary Risk Reduction (NRR) [A.2-6] is then the risk reduction that *must* be achieved by the Safety Functions in order to ensure that the Tolerable Risk is not exceeded, for a given level of EUC Risk.

Residual Risk [A.2-8] is the risk that is *actually* achieved for the specified hazardous events for the EUC / EUC Control System, but with the risk reduction afforded by the safety functions now taken into account[3].

## 2.3 Safety Integrity

We now need to address the question as to where Safety Integrity (i.e. "*the probability of a ... safety-related system satisfactorily performing the specified safety functions under all the stated conditions, within a stated period of time*" [A.2-11]) fits into the picture.

This is illustrated in Figure 2 and, whereas this diagram is not presented explicitly in IEC 61508, it follows logically from the above, as follows.

First of all, we have introduced the idea of a maximum possible amount of risk reduction, $\delta R(max)$ — relative to the EUC Risk ($R_{EUC}$) — which applies in the hypothetical case of a set of Safety Functions, implemented in safety-related systems (SRSs) and/or in other risk-reduction measures (ORRMs), which are entirely failure free. Clearly $\delta R(max)$ cannot be related to safety integrity since we have, in effect, assumed the latter property to be 100%. Therefore, the (hypothetical) maximum amount of achievable risk reduction must be dependent solely on what the safety functions do (i.e. their functionality) and on how well they do it (i.e. their performance); furthermore, the resulting minimum achievable risk ($R_{MA}$) is generally taken to be non-zero on the basis that there would always be some EUC hazardous situations that the safety functions would not be able to mitigate fully.

[3] Note that, in this illustration, we are in the fortunate position of achieving an actual level of overall risk reduction that is below that which is necessary for a safe state to be achieved overall.

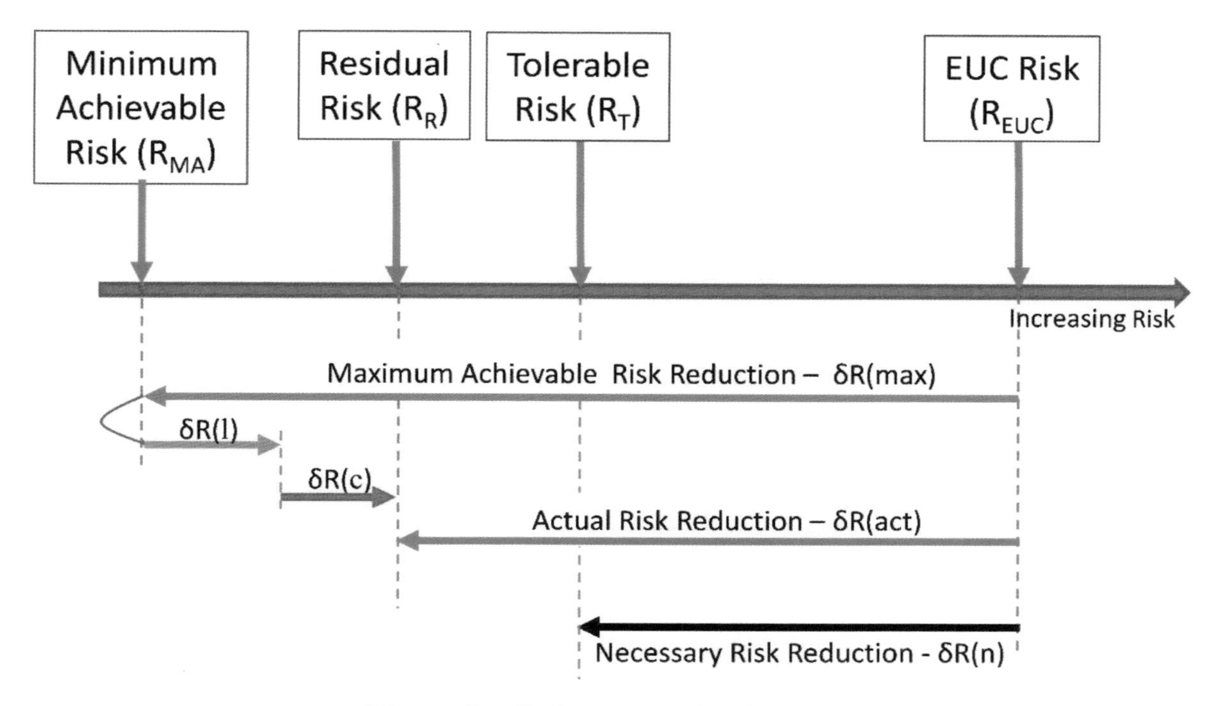


**Figure 2 ~ Safety Integrity Concepts**

Then we have δR (l), which represents an increase in risk arising from 'loss-type' failures of one or more safety functions; it is not a new risk as such — rather, it is effectively a shortening of the maximum achievable risk reduction arrow (i.e. a reduction below δR(max)).

δR(c), on the other hand, is a new source of risk because it arises from corrupt behaviour – i.e. incorrect or spurious operation — of one or more safety functions. Hence the Actual Risk Reduction is given by:

$$\delta R(act) = \delta R(max) - \delta R(c) - \delta R(l) \qquad \text{.................(1)}$$

The inescapable conclusion from Equation (1) is that, in order for the residual risk ($R_R$) to be no higher than the tolerable risk ($R_T$), the tolerable maximum rates of loss and corruption failures for the safety functions cannot be determined in isolation from the amount of necessary risk reduction (δR(n)) that the safety functions are required to provide in the first place.

In other words, carrying out a safety analysis of safety-function failures cannot, in itself, tell us if a state of tolerable risk would exist for the system as a whole (i.e. the EUC, EUC Control System, and SRSs) and, in order to determine the safety integrity required of the safety functions, we must first assess, under assumed failure-free conditions, the potential minimum achievable risk ($R_{MA}$) that the safety functions could provide, in relation to the EUC Risk ($R_{EUC}$)[4].

## 2.4 Safety Integrity Levels and Safety Assurance

A safety integrity level (SIL) is defined by IEC 61508-4 as follows [A.2-12]:

[4] This conclusion has, as explained further in Fowler (2015), serious implications for safety-assessment methodologies that, in effect, focus almost entirely on the analysis of safety function (or, at a lower-level SRS) failure — including in the Air Traffic Management and Rail sectors.

*"...discrete level (one out of a possible four), corresponding to a range of safety integrity values, where safety integrity level 4 has the highest level of safety integrity and safety integrity level 1 has the lowest".*

Note 3A to this definition emphasises that a SIL "*is **not** a property of a system, subsystem, element or component*". This begs the questions as to what SILs actually are, and what they are for; to answer these, we next need to consider some basic principles of safety assurance.

IEC 61508 does not provide an explicit definition of safety assurance although, as we will see, it is certainly implicit in many parts of the Standard.

A useful definition of safety assurance, derived from European law (Commission Regulation (EU) No 1035/2011), is as follows (European Commission 2011):

[A compelling argument supported by the body of evidence resulting from the application of*]*[5] "*all planned and systematic actions necessary to afford adequate confidence that a product, a service, an organisation or a functional system achieves acceptable, or tolerable, safety".*

"Confidence" is the operative word here and SILs play an important part in giving that confidence, to a level *appropriate to the risks involved*, by:

- setting standards for the design of products used in Safety-related Systems; and
- ensuring that an appropriate level of rigour is applied to the processes followed throughout the development, manufacture, operation, and maintenance of Safety-related Systems.

The manner in which SILs are *applied* is largely covered by Parts 2 and 3 of IEC 61508 and related to Phase 10, *et seq*., of the lifecycle, i.e. outside the scope of this paper. What we *are* concerned with herein is the way in which SILs are derived in the first place, as follows.

## 2.5 Derivation of SILs

IEC 61508 "*does not specify the safety integrity level requirements for any safety function, nor does it mandate how the SIL is determined*[6]". Nevertheless, we can set out some broad principles by using a fault tree representation of risk, as shown in Figure 3.

The fault tree is simply an alternative way of presenting the information in Figure 2, which is itself a refinement of Figure 1 and, hence, of Figure A.1 of IEC 61508-5 (IEC 2010).

In mathematical terms, the simple fault tree applies to a single Safety Function, in what IEC 61508-4 defines as "*a low demand mode of operation*". However, the author's intention here is not to detail a quantitative approach to SIL derivation — rather, it is to present the general logical relationships involved, from which we can deduce some sound principles for such an approach.

[5] The lead-in phrase has been added by the author to emphasise that assurance is an evidence-based activity and most of the "confidence" comes not from the actions *per se* but from the results thereof.

[6] See Note 2 to the Introduction of IEC 61508-4.

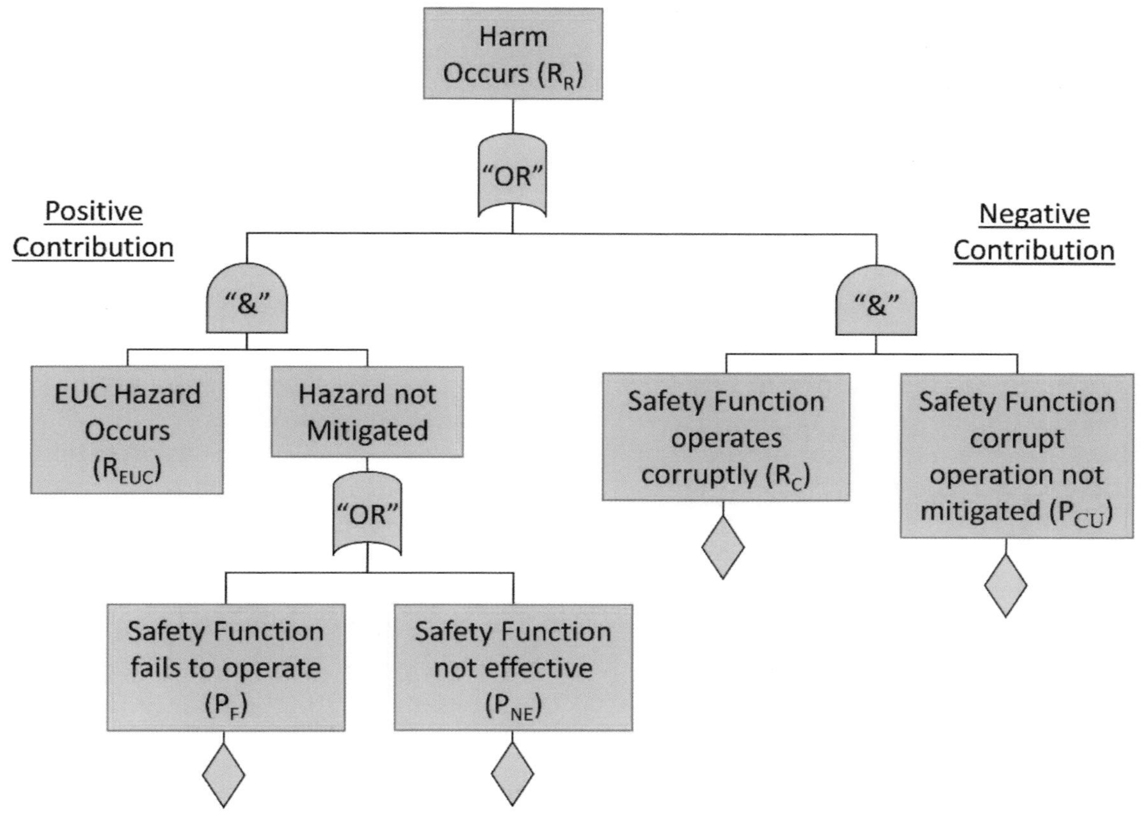


**Figure 3 ~ Fault Tree View**

In a general sense, a harmful event occurs if either:

- a hazard from the EUC occurs and is not mitigated by the Safety Function. Such a condition would occur if the Safety Function operated but was ineffective in mitigating the consequences of the EUC hazard under particular circumstances, or failed to operate at all; or
- the Safety Function operated corruptly, and the effects of this operation were not mitigated. Such a condition would occur if the safety function operated at the wrong time, under the wrong conditions, and/or in the wrong way, and if, for example, the corruption went undetected[7].

It is important to note again that the first of the above conditions is, in effect, the risk caused by failure to mitigate the EUC Risk, whereas the second condition is, in effect, a *new* risk that is not related to the EUC Risk.

As with most fault trees, Figure 3 can be used in a number of ways but if we were to *estimate* values (of probability or frequency, as appropriate) for the bottom-level events on each branch, *and* for the severity of the EUC hazard[8], we would end up with an estimate of risk for the top-level, harmful event.

Hence, relating Figure 3 back to Figure 2, we can see that:

---

[7] In neither case has the mitigating effect of Providence. i.e. the pure chance that a Hazardous Event or Hazardous Situation does *not* lead to a Harmful Event, been included in this simple model. In aviation for example, where the trajectory of an aircraft in flight has three degrees of freedom, Providence usually makes a major contribution to mitigating the consequences of a hazard; in the road and rail transport sectors, fewer degrees of freedom usually mean a lower contribution from Providence.

[8] If we wanted to model the likelihood of a range of hazard-severity outcomes, one way of so doing would be to connect the top-level risk to an Event Tree. That level of detail is, however, not necessary for the purposes of this paper.

1. in general, the top-level risk would equate to the residual risk ($R_R$);
2. if we were to set $P_F$ to 1.0 (i.e. a 100% probability that the safety function would have failed totally), then the top-level risk would become the EUC Risk ($R_U$)[9];
3. the difference between items 1 and 2 would equate to the actual risk reduction ($\delta R(max)$);
4. if $R_C$ and $P_F$ were set to zero, then the top-level risk would become the minimum achievable risk ($R_{MA}$); and
5. the difference between items 1 and 4 would equate to the sum of the risk due to the event "safety function failure to operate" ($\delta R(l)$) and the risk from "safety function corrupt operation" ($\delta R(c)$).

So, returning to the question of how to derive a SIL for the safety function, and referring to Figure 3, we know that EUC Risk ($R_{EUC}$) is (by definition) the risk which pertains when *no* safety function is in place, and can be represented by setting $P_F$, i.e. the probability of failure on demand for the safety function, to a value of unity on the fault tree. Note that, by setting $P_F$ to unity, we would effectively render the other base events irrelevant and, therefore, we would *not* need to know anything about $P_{NE}$, $R_C$ or $P_{CU}$ for this purpose. Given a reasonably accurate estimate for $R_{EUC}$, it would then be reasonably straightforward to compute a value for $P_F$ that would be required in order to *reduce* the top-level risk ($R_R$), from the level of the unmitigated EUC Risk, down to a tolerable level. That said, there are two important caveats to note here, as follows.

Firstly, for any value of $P_F$ *other* than unity, the other base events are not irrelevant, leaving us with two choices:

- to try to estimate values for $P_{NE}$, $R_C$ and $P_{CU}$; or
- redefine the computed value of $P_F$ such that it subsumes, into a single measure, the likelihood of all three possible causes of failure of the safety function — i.e. total loss, ineffectiveness (under specific circumstances), and corrupt behaviour.

The former option would probably not be viable since it would require considerably more detail about the safety function than would normally be available at this stage.

Fortunately, it turns out that the value of $P_F$ defined as above is synonymous with what IEC 61508 terms the "*target failure measure*" [A.2-13] for the safety function, and Table [10] below shows the four SIL levels together with their corresponding bands of target failure measures, for what IEC 61508-4 defines respectively as "*high demand*" and "*low-demand*" modes of operation.

Therefore, it would be a simple matter to look up the estimated value for $P_F$ in Table 1 – in this case the third column[11] – and read off the corresponding SIL.

Secondly, estimating a value for EUC Risk can be quite difficult for any safety-related application, not the least for the transport sector. However, IEC 61508-1, Sub-section 7.5.2.4 (IEC 2010) allows for an alternative way of deriving target failure measures that does *not* depend on direct knowledge of the unmitigated EUC Risk – this is discussed further in Phases 4 and 5 below.

[9] Values for $F_C$ and $P_{NE}$ are not relevant here because the of EUC Risk is defined as being the risk pertaining in the complete absence of the safety function – this is represented fully by simply setting $P_F$ to unity.

[10] **Table** herein simply combines Tables 2 and 3 of Sub-section 7.6.2.9 of IEC 61508-1 (IEC 2010).

[11] That is because Figure 3 applies specifically to a "low-demand" case; however, the same principle would apply to a high-demand situation.

**Table 1 ~ Target Failure Measures & Safety Integrity Levels (SILs)**

| SIL | Average frequency of a dangerous failure of the safety function per hour | Average probability of a dangerous failure on demand of the safety function |
|---|---|---|
| 4 | $\geq 10^{-9}$ to $< 10^{-8}$ | $\geq 10^{-5}$ to $< 10^{-4}$ |
| 3 | $\geq 10^{-8}$ to $< 10^{-7}$ | $\geq 10^{-4}$ to $< 10^{-3}$ |
| 2 | $\geq 10^{-7}$ to $< 10^{-6}$ | $\geq 10^{-3}$ to $< 10^{-2}$ |
| 1 | $\geq 10^{-6}$ to $< 10^{-5}$ | $\geq 10^{-2}$ to $< 10^{-1}$ |

# 3 IEC 61508 Lifecycle Processes

## 3.1 Overview

This section explains how the above IEC 61508 principles are applied throughout the related phases of the IEC 61508 lifecycle.

Figure 4 shows the *overall* processes involved in the specification and realisation of the safety properties required by the SRSs and ORRMs in order that a tolerable level of risk could be achieved for the EUC / EUC Control System.

The diagram is based on Figure 2 of IEC 61508-1 (IEC 2010), with the following modifications:

- items in grey are shown for context reasons only, and fall outside the scope of this paper;
- Phases 6 to 8 have been omitted entirely as they cover only the planning for Phases 12 to 14 respectively;
- a summary of the main outputs of each relevant phase has been added to the diagram;
- the specification of safety requirements for ORRMs, i.e. Phase 11, falls within the scope of this paper, even though it is outside the scope of IEC 61508 itself; and
- IEC 61508's use of the term "*E/E/PE (System) — Electrical/Electronic/Programmable Electronic (System)*" was felt to be too specific and limiting for the purposes of this paper; therefore, the more general term "safety-related system (SRS)" is used instead herein so as to allow human and procedural elements to be included as well as (and possibly instead of) technical equipment.

An outline of the various pertinent lifecycle phases is set out as follows. Where applicable, the relevant IEC 61508-1 sub-section number is given and, as far as possible, the text is taken directly from the Standard. However, in some cases it has been found necessary to modify, or add to, the 61508-1 text in order to clarify a particular point, or to cater for some of the additional complexities of the transport sector; wherever this is the case, the rationale for such changes is given.

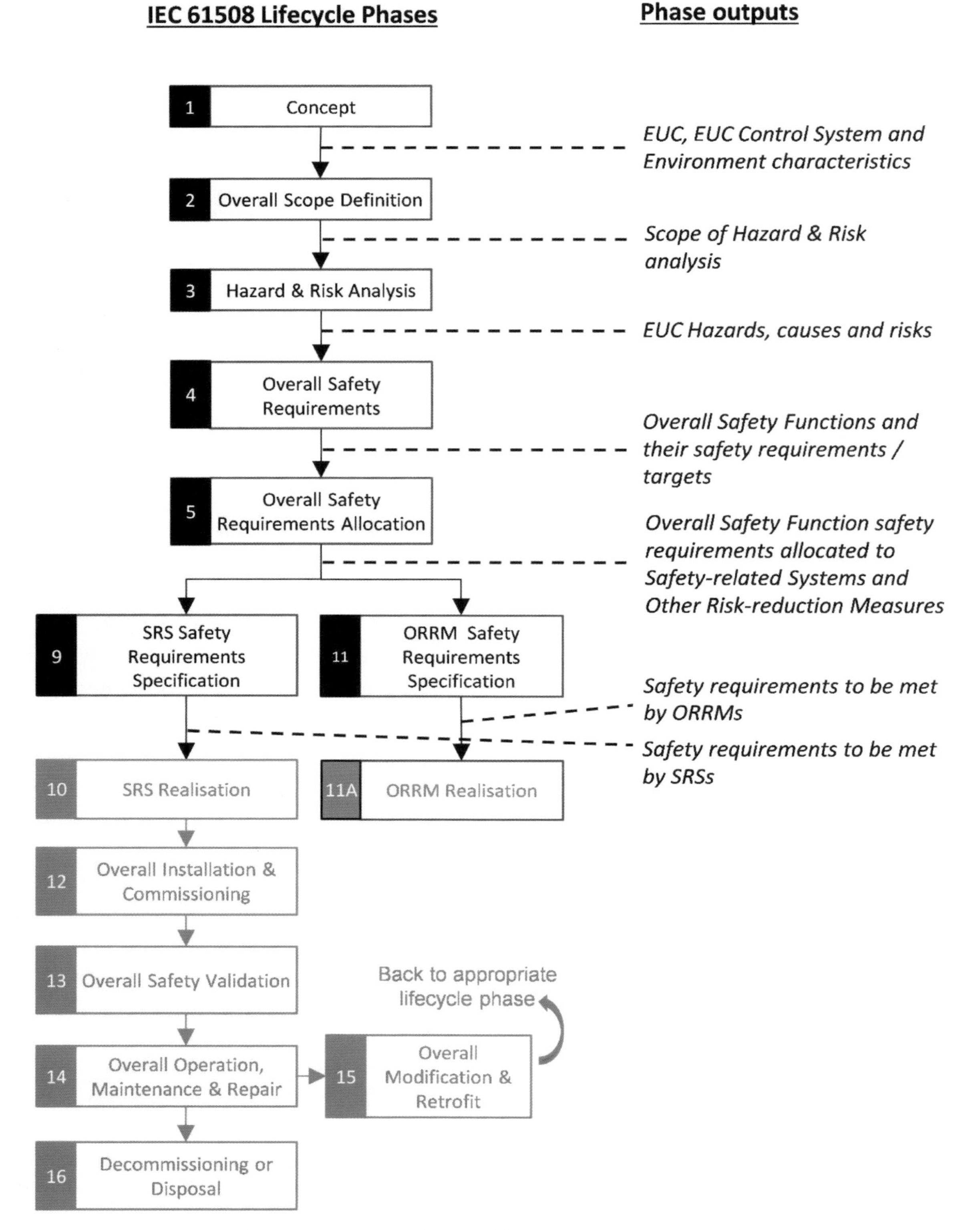


**Figure 4 ~ IEC 61508 Overall Safety Lifecycle**

For each lifecycle phase, guidance is given on the key issues involved in meeting the stated aim. Where applicable, this guidance is illustrated by considering a very simple, hypothetical road-transport application — the safety of pedestrians crossing a busy road[12]; however, unlike in subsequent parts of this paper, which will include more-detailed, worked examples, the discussion below is limited to highlighting the general issues involved in each phase, without seeking to resolve them for the specific case.

## 3.2 Phase 1: Concept (IEC 61508-1, Sub-section 7.2)

### *3.2.1 Aim*

The aim of this phase is to gather as much information about the EUC / EUC Control System and its environment as is necessary and sufficient to enable the other safety-lifecycle activities to be satisfactorily carried out.

It is important to note that, as an enabling activity, this would be a precursor to, but not form part of, the safety assessment *per se* and would require *substantial* operational and system-engineering specialist input, relevant to each specific application.

### *3.2.2 Guidance*

Section 4 of Fowler and Pierce (2012) explains the importance of the relationships involved here, in terms of good requirements-engineering practice. In essence, if we are to derive a valid set of safety requirements — initially for overall safety functions and, from those, for the corresponding SRSs and ORRMs — we need to establish precisely those properties of the EUC / EUC Control System and environment that could impact on the overall safety functions and their ability to mitigate EUC Risk to the degree necessary.

To get the Phase 1 process underway, the first step would have to be to decide what the EUC / EUC Control System actually comprises — something that is not necessarily immediately obvious! For our pedestrian-safety problem, we might decide to define the EUC as the flow of road traffic in the area of potential conflict with pedestrians; the flow of pedestrians would then form part of the *operating* environment. The logic of that decision is that it is the road traffic that presents hazards to the pedestrians, *not* vice-versa.

Properties of the traffic that we would need to know would include the various types of vehicle using the road (cars, buses, vans, heavy goods vehicles, automated vehicles, cyclists, etc.), the relative numbers of each, and their flow rates at different times of the day or week.

Examples of properties of the environment are:

- the road layout (e.g. road width, visibility, number of traffic lanes, single or dual carriageway, one- or two-way traffic, and the width available for the passage of pedestrians), and typical weather conditions (*physical* environment);
- characteristics of any proximate facilities (schools, hospitals, transport hubs, etc.) that could affect the flow of pedestrians at different times of the day or week (*physical* environment);

[12] This is *purely* for the purposes of illustration. It is not suggested that that the safety assessment of a proposed pedestrian crossing would have to be carried out this way.

- pedestrian characteristics (age, capability / disability, etc. and the relative numbers of each), pedestrian behaviours (e.g. inclination to avoid being hit by vehicles and yet a tendency to take risks), and the crossing rate of pedestrians (*operating* environment);
- vehicle speed limits, rights of way, etc. (*legal* environment);
- potential roadworks (*maintenance* environment).

It is important to note here that there is an extant EUC Control System — i.e. the natural inclination (if not legal obligation!) of vehicle drivers to comply with traffic signs and signals, and to avoid collisions with pedestrians and other vehicles in the flow. Indeed, in many road environments, with low density of pedestrians and road traffic, the EUC Control System would probably be sufficient in itself to achieve a tolerable level of risk of harm to pedestrians.

For more complex applications, e.g. ATM and rail systems, the EUC, EUC Control System, and environment would not only themselves be far more complex than the above and would, therefore, require much more detailed descriptions and analyses, but would also play a much greater part in the achievement of tolerable risk for the EUC (see the guidance on overall safety-function SIRs in Sub-section 3.5.2 below)[13].

## 3.3 Phase 2: Overall Scope Definition (IEC 61508-1, Sub-section 7.3)

### *3.3.1 Aim*

The aim of this phase is to define the scope of the hazard and risk analysis, for Phase 3.

### *3.3.2 Guidance*

Sub-section 7.3 of IEC 61508-1 starts this phase by determining the boundary of the EUC and the EUC Control System. The important issue here is that we are not trying to determine the actual limits of the EUC / EUC Control System, or of its environment, in absolute terms, since that should have been done in Phase 1 — rather, the requirement here is to decide the boundaries purely of the analysis work, in terms of the EUC / EUC Control System and its environment, as appropriate to the overall safety functions(s) that we are wanting to specify.

For example, for the hazard and risk analysis associated with our pedestrian-safety problem, we might want to:
- limit the hazards for pedestrians to those occurring only within a designated crossing area;
- exclude hazards not caused by the traffic; and
- exclude vehicle-vehicle collisions and vehicle incursions into pedestrian-only areas.

IEC 61508-1, Sub-section 7.3.1 helpfully notes that "*Several iterations between the overall scope definition and the hazard and risk analysis* [Phase 3] *may be necessary*"!

---

[13] That will also become increasingly true in road transport with the advent of automated vehicles.

## 3.4 Phase 3: Hazard and Risk Analysis (IEC 61508-1, Sub-section 7.4)

### *3.4.1 Aim*

The aim of this phase is to determine and characterise all the hazards and risks associated with the EUC / EUC Control System in the stated operational environment, within the scope identified in Phase 2.

### *3.4.2 Guidance*

It is vital to understand that, at this stage, the *subject* of the hazard and risk analysis is the EUC and its control system, *not* the overall safety function(s) or SRSs, which come later in the process. IEC 61508-1, Sub-section 7.4.1, Note 1, stresses this point as follows:

> *"This subclause is necessary in order that the safety requirements for the ... safety-related systems are based on a systematic risk-based approach. This cannot be done unless the EUC and the EUC Control System are considered".*

Regrettably, this fundamental point does not seem to be widely understood — at least not in some key areas of the transport sector — as discussed in Fowler (2015).

Sub-section 7.4.1 of IEC 61508-1 identifies the following three distinct steps in the hazard and risk analysis process.

**Firstly:** to determine the hazards, hazardous events and hazardous situations relating to the EUC and the EUC Control System, in all modes of operation, and for all reasonably foreseeable circumstances — principally normal, abnormal, and failure conditions.

This requires that the hazards, hazardous events, and hazardous situations of the EUC and the EUC Control System be determined under all reasonably foreseeable circumstances, including fault conditions, reasonably foreseeable misuse, and malevolent or unauthorised action. It must include all relevant human factor issues and must give particular attention to abnormal, or infrequent, modes of operation of the EUC.

The main EUC hazard of interest for our pedestrian-safety problem is probably the *hazardous situation* that:

> *Haz #1: the respective needs of the traffic flow (EUC / EUC Control System) and the flow of pedestrians result in the moving traffic and pedestrians intending to occupy the same area of the road surface at the same time.*

It might also be relevant to consider two additional hazards, which could arise from failure of the EUC Control System and in the environment, respectively:

> *Haz #2: failure of a vehicle driver to take action to avoid a stray pedestrian;*
> *Haz #3: failure of a pedestrian to notice an on-coming vehicle.*

Specific IEC 61508 requirements on EUC Control System failures are explained in Phase 5 (Sub-section 3.6.2) below.

**Secondly:** to determine the preconditions and sequences leading to the hazardous events and hazardous situations.

This step is important in that knowledge of what causes or leads to hazardous events or situations can itself:

- lead to a clearer understanding of how such events or situations could be avoided — i.e. the hazard eliminated — sometimes the preferred solution; and/or
- facilitate estimation (in step 3) of how frequently such events or situations would be likely to occur.

Automatic traffic counters would be a typical means of collecting overall road-usage data on the vehicle flows around the area of interest for our pedestrian-safety problem. These could be supplemented by more selective techniques such as on-site enumerators or video cameras to collect pedestrian and cycle-survey data. It would also be useful to know:

- what determines the patterns of vehicle flows including what percentages of traffic is simply transiting the local area as opposed to requiring access to local facilities;
- about the presence of local facilities that could have a particular impact on pedestrian and/or vehicle flows, i.e. schools, workplaces, shopping centres, bus stops, etc.

**Thirdly:** to determine either the EUC Risks associated with the hazardous events and hazardous situations or the consequences of those events / situations.

This step might seem to be an essential condition for determining the NRR and thence the SIL and safety integrity requirements (SIRs), for each overall safety function. However, we will see in Phase 4 that IEC 61508-1 provides a means of avoiding the necessity of having to estimate EUC Risk, provided the consequences of the EUC hazards are known; we have, therefore, extended the original text of IEC 61508-1, Sub-section 7.4.1.3 to allow for this option.

It is not difficult to see that the estimation of EUC Risk depends heavily on information gathered in the preceding phases and processes, not the least of which is a thorough and complete understanding, from Phase 1, of the EUC, its control system, and its environment.

## 3.5 Phase 4: Overall Safety Requirements (IEC 61508-1, Sub-section 7.5)

### *3.5.1 Aim*

The aim of this phase is to produce a specification of the *overall safety requirements* — i.e. in terms of functional safety requirements and safety integrity requirements — for the overall safety function(s) in order to achieve the required level of functional safety.

### *3.5.2 Guidance*

The language of Sub-section 7.5 (and as we will see below, of Sub-section 7.6) of IEC 61508-1 can be quite difficult in places. In order to address this, it was decided to:

- use the term "functional safety requirements" (FSRs) herein, instead of paragraph 7.5.1's "*safety functions* [sic] *requirements*"; and
- use the term "overall safety function(s)" instead of paragraph 7.5.1's "E*/E/PE safety-related systems and other risk reduction measures*", since we need to specify only the *overall* safety functions at this stage, not SRSs or ORRMs as they will be done later, in Phase 5.

Three process steps can be discerned from IEC 61508-1, Sub-section 7.5, as follows:

**Firstly:** to identify a set of overall safety functions, based on the EUC hazardous events derived from the hazard and risk analysis of Phase 3.

In the context of this step, Note 1 of IEC 61508-1, Sub-section 7.5.2.1, clarifies that: "*It will be necessary to create an overall safety function for each hazardous event*" associated with the EUC.

The importance of the word "overall" here is that the safety specification at this level is intended to be independent of whether the safety function would be realised as an SRS or ORRM (or both).

**Secondly:** to determine the required functional properties, i.e. the FSRs, of each overall safety function so as to address the related EUC hazard.

One way of achieving this is to start with wording of the related hazard and turn it into a high-level functional requirement statement of what needs to be done in order to mitigate the hazard — but *not* how this is to be done.

Taking our pedestrian-safety problem as an example, there are three quite different ways in which we could reduce EUC Risk — viz mitigate the hazard consequences, reduce the frequency of occurrence of the hazard, or remove the hazard completely — with markedly different realisations such as a controlled or uncontrolled pedestrian crossing, a road bypass, a footbridge or an underpass.

The main hazard (Haz #1 above) was about "*...moving traffic and pedestrians intending to occupy the same area of the road surface at the same time*". All we need to do, in order to remove the hazard, would be to negate at least one of the two conditions that define the hazard, so that we end up with the following simple functional requirement statement:

> *"FSR 1: In the area of potential conflict, the overall safety function shall ensure the safe separation of pedestrians and moving road traffic temporally and/or spatially".*

Such a requirement does not preclude any particular mitigation strategy or ultimate technological solution and yet is, at the highest level, sufficient in itself provided the associated required safety integrity requirements (SIRs) accompany it.

**Thirdly:** to determine the SIRs for each overall safety function so as to achieve a tolerable level of risk.

Sub-section 7.5.2.4 of IEC 61508-1 explains that the overall SIRs must be specified in terms of either:

- "*the risk reduction required to achieve the tolerable risk*"; or
- "*the tolerable* [EUC] *hazardous event rate so as to meet the tolerable risk*".

Note that this means that the SIRs at this level are not properties of the safety function to which they relate — whereas the overall functional safety requirements, derived in the previous step, specify what the overall safety function has to do (functionally), the SIRs for the overall safety function specify a target amount of EUC Risk *reduction* that the safety function has to provide in order to achieve a tolerable level of risk overall.

The latter of the above two bulleted options is a less-direct way of expressing required risk reduction but, as discussed in Sub-section 2.5 herein, is more pragmatic — especially in the transport sector. In order to determine the tolerable rate of occurrence of an EUC hazardous event, to meet the tolerable risk, we would need to know the consequences of the EUC hazardous event, in terms of the probability that occurrence of the event would result in the harm to which the tolerable risk relates. This could be done quantitatively or qualitatively, aided by, for example, some form of hazard severity matrix / risk classification scheme that had been pre-defined for the particular application *and* for use at the EUC-hazard level; IEC 61508-5 provides general advice on a wide range of such techniques.

Further, in relation to the derivation of overall SIRs, where any reliance is placed on a possible contribution to achievement of tolerable risk that might be assumed (or actually specified) for the EUC Control System, it is also very important to note the provisions of Sub-section 7.5.2.5 of IEC 61508-1, which state that:

*"If, in assessing the EUC Risk, the average frequency of dangerous failures of a single EUC control system function is claimed as being lower than $10^{-5}$ dangerous failures per hour then the EUC Control System shall* [itself also] *be considered to be a safety-related control system* [and] *subject to the requirements of this Standard".*

The Note to Sub-section 7.5.2.5 of IEC 61508-1 then goes on to clarify that whatever average frequency of dangerous failures is claimed for the EUC Control System function (i.e. if below $10^{-5}$ per hour), all the IEC 61508 requirements appropriate to the corresponding SIL would need to be met for the EUC Control System.

Finally of note is Sub-section 7.5.2.6, which discusses the case where failures of the EUC Control System place a demand on one or more SRSs and/or ORRMs, but where the intention is not to designate the EUC Control System as an SRS; it sets out a number of assurance requirements regarding the provision of data to support the rate of dangerous failure claimed for the EUC Control System, which in any event must not be lower than $10^{-5}$ dangerous failures per hour.

For our pedestrian-safety problem, we have already identified the EUC Control System as being the actions of pedestrians and vehicles to avoid each other, and that in areas of low pedestrian and traffic flows this can provide an adequate level of safety. What we also know is that this particular EUC Control System is non-linear, and its effectiveness diminishes substantially as the pedestrian and traffic flow rates increase. Therefore, great care would need to be taken in making any claims in regard to its safety properties.

## 3.6 Phase 5: Overall Safety Requirements Allocation (IEC 61508-1, Sub-section 7.6)

### *3.6.1 Aim*

The aim of this phase is to allocate to SRS(s) and/or ORRM(s), the functional safety requirements and safety integrity requirements, which were derived for the corresponding overall safety function in Phase 4.

### *3.6.2 Guidance*

The language of IEC 61508-1 is again slightly problematic in that it does not always distinguish between "safety function" and "overall safety function" (e.g. in Sub-section 7.6.2.9) and the relationship between safety integrity requirements, target failure measures. and SILs is not always as clear as it might be.

This author's understanding is that:

- the *safety integrity requirements* for each of the *overall safety functions*, which are being allocated to SRSs / ORRMs in this phase, are as described in Sub-section 3.5.2 herein, i.e. they are either the risk reduction, or the tolerable EUC-hazard occurrence rate, to be achieved by the associated overall safety function;
- a *target failure measure* is the specific, inclusive safety-integrity property required of each SRS in order to satisfy the allocated *overall safety integrity requirement*[14], and from which the SIL for the safety function can be derived. It is specified as either: the average probability of a dangerous failure on demand of the safety function, for a low-demand mode of operation, *or* the average frequency of a dangerous failure of the safety

[14] See also the discussion on the Derivation of SILs in Sub-section 2.5 herein

function for a high-demand (or a continuous) mode of operation (IEC 61508-1, Sub-section 7.6.2.5); and
- the allocation of requirements from an overall safety function to SRSs could be done at the whole-SRS level or at the level of the constituent safety functions within the SRS.

Three process steps can be discerned from IEC 61508-1, Sub-section 7.6, as follows:

**Firstly:** to decide how the overall safety function for each hazard is to be implemented in terms of SRS(s) or ORRM(s) or a combination of the two, as appropriate. This is a relatively straightforward requirements-allocation process and the first of many further steps towards defining a solution to the top-level requirements of the overall safety function.

For our pedestrian-safety problem, we appear to have at least three main options:
- to install some form of pedestrian-control crossing in order to allow pedestrians and moving traffic to share the same road space but separate them temporally;
- to build a footbridge (or underpass) in order to separate pedestrians from traffic spatially; or
- restrict the volume of traffic using the road in question, providing a bypass for through traffic if necessary.

The first option is clearly an SRS but supported by ORRMs in the form of road markings, warning signs and possibly barriers; the second and third options would clearly be ORRMs since they are non-functional in nature.

It would make sense to introduce the *As Low As Reasonably Practicable,* "ALARP", principle — see IEC 61508-5 and HSE (2021) — at this stage in order to decide which of the above options should be adopted, since both the costs involved, and the risk reduction achievable, would probably be significantly different. This could be done qualitatively or quantitatively, albeit the latter would probably require the use of a socially acceptable Value of a Prevented Fatality (HSE 2018).

**Secondly:** to allocate the functional requirements contained in the specification for the overall safety function (i.e. the overall FSRs) to the designated SRS(s) and/or ORRM(s).

This is also relatively straightforward except that an important question at this stage might be whether more details could and should be decided for the options, e.g. what sort of pedestrian crossing (Zebra, Pelican, Puffin, Toucan, *et alia*) would best meet the functional and integrity requirements of the overall safety function, given the properties of the operational environment concerned?

**Thirdly:** to allocate the safety integrity requirements contained in the specification for the overall safety function (i.e. the overall SIRs) to the designated SRS(s) and/or its constituent safety functions, thence derive a *target failure measure* and an associated SIL for each SRS / constituent safety functions.

IEC 61508-1 deliberately does not allocate overall SIRs to (nor derive target failure measures nor SILs for) ORRMs; however, it is likely that some equivalent to SILs would be necessary for ORRMs in some transport applications.

Allocation of overall SIRs to SRSs is not straightforward and the advice of IEC 61508-1, Sub-section 7.6.2.6, that the "*allocation of the safety integrity requirements shall be carried out using appropriate techniques for the combination of probabilities*" is less than helpful! The problem is that, at this level of abstraction, there is no technological basis for apportioning 'probabilities' between SRSs belonging to the same overall safety function.

In terms of SIL derivation, we have the additional problem that the lack of knowledge of their ultimate technological implementation also means that that it would not really be possible to show independence between SRSs (and/or between the constituent safety functions of an SRS) and, therefore, in line with IEC 61508-1, Sub-section 7.6.2.10, it would be necessary to assign to all SRSs / safety function the same SIL — i.e. the highest SIL of all the SRSs belonging to the same overall safety function[15].

It follows, therefore, that whereas there is some guidance on the processes required under this step in the discussion on the derivation of SILs, in Sub-section 2.5 herein, any results must be regarded as tentative and subject to confirmation at Phase 10, (SRS realisation) of the lifecycle. Alternatively, IEC 61508-1, Sub-section 7.5.2.3, has two helpful notes:

- Note 1, which states specifically that, "*some of the qualitative methods used to determine SILs in IEC 61508-5, Annexes E and F, progress directly from the risk parameters to the safety integrity levels — hence, in such cases, the overall SIRs are implicitly rather than explicitly stated because they are "incorporated in the method itself*"; and
- Note 5, which allows more-generally for situations where an application sector international standard exists that includes appropriate methods for directly determining the safety integrity requirements; it may be used to meet the requirements of this part of the Standard.

## 3.7 Phase 9: SRS Safety Requirements Specification (IEC 61508-1, Sub-section 7.10)

[Note that Phases 6 to 8 of IEC 61508-1 fall outside the scope of this paper]

### *3.7.1 Aim*

The aim of this phase is to develop further the safety requirements for the SRS identified in Phase 5, in terms of its FSRs and the SIRs, in order to achieve the required functional safety.

### *3.7.2 Guidance*

The above "aim" has been derived from the "objective" of IEC 61508-1, Sub-section 7.10.1, and is consistent with Sub-section 7.10.2.1, which states clearly that:

> *"The* [SRS] *safety requirements specification shall be derived from the allocation of safety requirements specified in* [Phase 5] *... ".*

However, the phraseology has been altered somewhat, to remove apparent ambiguities, e.g. "*safety functions requirements*" becomes "*functional safety requirements*", i.e. FSRs. Furthermore, the term "safety function" is never used generically herein — instead, the formal IEC 61508-4 definitions, at Appendix A hereto, have been adhered to and are understood as follows:

- an "*overall safety function*" is the highest level of abstraction of the single set of SRSs and/or ORRMs that provides the complete response to a specific EUC Hazard; and
- a "*safety function*" is one of a number of functional entities implemented by an SRS.

The following, slightly-paraphrased, note from Sub-section 7.10.2.2 of IEC 61508-1, explains the purpose of the outputs of this phase of the lifecycle and their important relationship with the requirements derived in the previous two phases:

---

[15] IEC 61508-1, Sub-section 7.6.2.10 is actually about independence between safety functions *within* SRSs; however, the same reasoning could, and should, be applied also to independence *between* SRSs.

> *"Note: The objective is to describe, in terms not specific to the equipment, the* [required safety properties of the SRS(s)]. *The* [SRS Safety Requirements] *Specification can then be verified against the outputs of the 'overall safety requirements' and the 'overall safety requirements allocation' phases, and used as a basis of the realisation of the* [SRS]. *Equipment designers can use the Specification as a basis for selecting the equipment and architecture".*

Unfortunately, Sub-section 7.10.2.2 itself is less helpful in that it states that:

> *"The SRS safety requirements specification shall contain requirements for the safety functions and their associated safety integrity levels".*

Since SILs might already been allocated to the safety functions implemented by SRSs, in Phase 5[16], and the fact that — in any event — SILs are not system properties (see Note 3A to the definition of a SIL [A.2-12]), the broader term "*safety integrity requirements*", is used instead of the more restrictive term "*safety integrity levels*".

The following two process steps can be discerned from IEC 6150-1, Sub-section 7.10:

**Firstly:** to derive a full description of the SRS's required functional and performance properties — and its behaviour in relation to the EUC, EUC Control System and environment — that are necessary to achieve the required risk reduction (i.e. NRR).

Figure 2 herein shows that this reduction needs to be greater than the NRR in order to allow for the risk associated with loss failure of the safety function ($\delta R(l)$) and the risk associated with corrupt operation of the safety function ($\delta R(c)$). Therefore, the resulting SRS Safety Requirements Specification will need to be supported by analysis to show that the SRS is able to provide the required risk reduction under all normal, abnormal and failure states of the EUC, the EUC control system and the environment, as well as the transitions between those states.

Sub-section 7.10.2.6 of IEC 61508-1, in particular, places great emphasis on the need for a description of the workings of the SRS at a functional level, including:

- a description of all the safety functions, how they work together to achieve the required functional safety and whether they operate in low-demand, high-demand or continuous modes of operation;
- the required performance attributes of each safety function, e.g. timing properties and, for more data-intensive applications than possibly envisaged by IEC 61508, data accuracy, latency, refresh rate, and overload tolerance;
- all interfaces[17] that are necessary to achieve the required functional safety;
- all relevant modes of operation of the EUC;
- response of the SRSs to abnormal conditions that might arise in the EUC or its environment;
- all required modes of behaviour of the SRSs — in particular, its failure behaviour and the required response in the event of such failure.

The underlying point is clear: before we get into the SIRs for the SRS, we need to fully demonstrate the adequacy of the *FSRs* in satisfying the requirements for EUC-risk reduction, in the absence of failure of the SRS.

[16] That does not mean that SIL derivation could not, if necessary, be repeated at this level of the requirements hierarchy; rather, the point is that it is not the most important consideration at this stage in the process.

[17] IEC 61508-1 also includes "operator" interfaces at this level. We have excluded that on the ground that human operators can by definition form part of a physical SRS and should be left until the physical (or at least logical) design stage of the development lifecycle.

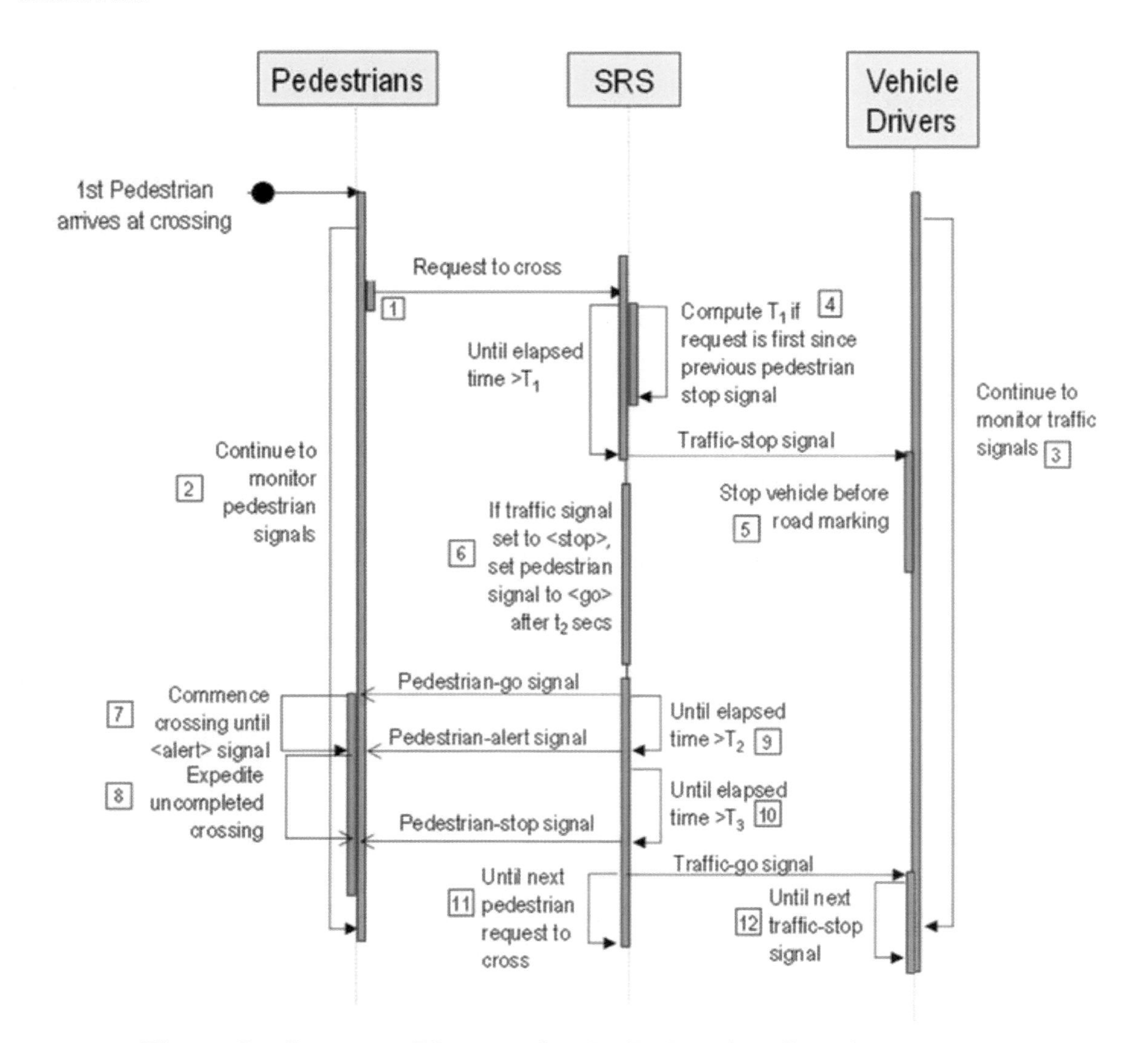


**Figure 5 ~ Sequence Diagram for the Pedestrian Crossing Example**

Assuming some form of temporal-separation solution and, given that the functionality is likely to be relatively straightforward, the FSRs specification for our pedestrian-crossing could be based initially on the sequence diagram (Sparx Systems 2020) shown in Figure 5, as the main means of defining the logic and relationships involved for normal operational conditions.

A sequence diagram shows object interactions arranged in time sequence, represented by the vertical dimension. It depicts the objects (identified across the top) involved in the scenario and the sequence of messages exchanged between the objects needed to carry out the functionality of the scenario.

Figure 5 represents a normal (or typical) scenario which starts when a pedestrian arrives at the crossing, during a period of road-traffic flow, and the sequence continues (for one iteration) as follows:

[1] the pedestrian initiates a request to cross;

[2] thereafter, pedestrians continue to monitor and comply with the pedestrian signals;

[3] vehicle drivers continue to monitor and comply with traffic signals;

[4] if the pedestrian request is the first request since the previous pedestrian <stop> signal, the SRS computes the time delay T1, else ignores the request;

[5] once the elapsed time since the pedestrian request to cross exceeds T1 then the SRS signals the traffic to <stop>;

[6] drivers stop vehicles before the appropriate road markings;

[7] after confirming that the traffic signal is set to <stop>, the SRS sets the pedestrian signal to <go> after a delay of T2 secs;

[8] pedestrians commence crossing until the crossing signal changes to <alert>;

[9] when the elapsed time since initiation of the <go> signal exceeds T3, the SRS sets the pedestrian signal to <alert>;

[10] at the <alert> signal, pedestrians already on the crossing complete their crossing expeditiously;

[11] when the elapsed time since pedestrian <alert> signal exceeds T4, SRS sets the pedestrian signal to <stop>;

[12] SRS maintains state until the next pedestrian request to cross;

[13] Traffic continues to flow until the next traffic <stop> signal.

The FSRs should capture the essence of the above scenario but in a more-formal language, starting with the overall functional requirement that pedestrian and traffic flows *shall* be controlled in turn such that pedestrians and moving traffic cannot occupy the designated crossing area at the same time. It should also include requirements that:

- call up the scenario itself as the required behaviour of the SRS;
- specify the time interval T1 (or its method of calculation) between the first crossing request by a pedestrian after traffic restart and the next instruction for the traffic to stop, such that there is a safe balance between the needs of pedestrians and the need to avoid excessive traffic queues;
- specify the total time interval T2 + T3 for which the traffic must be halted, such that it would be adequate to cater for the number and physical capabilities of pedestrians who might be using the crossing at the time;
- specify the time interval T3 such that pedestrians are given adequate warning of the imminent end of the crossing period;
- specify the means of indicating the state of the crossing to pedestrians such that it would be suitable for users with impaired hearing or vision;
- specify the default state of the crossing to be such that the traffic continues to flow in the absence of an input request from a pedestrian;
- specify that, under no circumstances, shall the pedestrian <go> signal and vehicle <go> signal states exist at the same time;
- specify the dimensions of the crossing area to be such that its pedestrian capacity would be great enough to handle the peak number of pedestrians under all reasonably foreseeable conditions.

The process of development of the FSRs would also need to address other, abnormal scenarios such as extreme weather and road maintenance which might affect the effectiveness of the SRS in providing the required risk reduction. Analysis should include assessment of the consequences and likely frequency of occurrence of such abnormal events so that account could be taken of the associated risk increase when developing the SIRs.

It is well worth noting here that sequence diagrams can be expanded, into greater detail, at lower levels in the system hierarchy — in the case of the pedestrian crossing shown in Figure 5, for example, by replacing the single "SRS" actor by the internal elements of its functional, logical and/or physical architecture — thus capturing the required behaviour at these levels, as required.

**Secondly:** to specify the detailed SIRs, for each safety function identified for each SRS, as follows:

- identification of the potential failure modes of the SRS(s), as well as the EUC control system;
- identification, and capture as additional FSRs, the possible mitigations of the frequency and/or consequences of those failures; and
- specification of the maximum occurrence rates of those failures, taking account of those mitigations, and of the minimum achievable risk that the SRS could provide, under assumed failure-free conditions[18], such that a tolerable level of risk for the EUC is achieved overall.

Thus far in the lifecycle, SIRs derivation has been largely deductive, "top-down", whereas it now needs to be inductive, "bottom-up". This is consistent with the note to the third step of Phase 4 (Sub-section 3.5.2 herein), that SIRs derived at the *overall safety function* level are *not* properties of the safety function itself.

It is also in line with commonly-accepted safety-assessment practices, and IEC 61508-5 suggests various techniques that could be used for this purpose — of course, any techniques and parameters involved must be demonstrably suitable for the application, and at the level in the requirements hierarchy, for which they are to be used.

It also needs to be understood that SIRs at this level would necessarily be based on an somewhat arbitrary apportionment of failure rates between the safety functions[19]; they would, therefore, need to be reviewed for the physical architecture of the SRS, at the realisation stage (Phase 10 in Figure 4) in order to avoid, for example, allocating inappropriately-low required failure rates to human operators.

### 3.8 Phase 10: Safety Requirements Specification for Other Risk-reduction Measures (IEC 61508-1, Sub-section 7.11)

IEC 61508 makes many references to ORRMs but otherwise rules them as outside the scope of the Standard — presumably as they are seen as being non-functional by nature. Therefore, since this paper is primarily about IEC 61508, albeit its potential application to the transport sector, it would not be relevant to introduce a lot of detail on ORRMs here.

Suffice it to say at this juncture that, whether they take the form of road, rail or runway layouts, or pre-defined routes through a block of airspace, ORRMs have a major role to play in the reduction of EUC Risk and, therefore, need to be treated, as far as possible, with the same (or equivalent) rigour as given to SRSs in IEC 61508. For example, in the case of a footbridge solution to our pedestrian-safety problem, it would be prudent to specify the relevant construction regulations.

[18] See Sub-section 2.3 herein

[19] The nature of risk classification schemes, for example, means that they spread the target risk evenly across all failure causes.

# 4 Conclusions

Previous research, e.g. Fowler (2015), showed that, in some areas of the transport sector in Europe, some safety assessment practices, e.g. that of EUROCONTROL (2015) for ATM, and CENELEC (1999) for rail, focussed too much on system reliability and not enough on system functionality, contrary to, *inter alia*, the most basic principles of the international functional-safety standard IEC 61508 (IEC 2010).

This paper, to be published as a series in three parts, sets out to show what functional safety assessments for transport applications might look like if they followed the safety principles and lifecycle processes set out in IEC 61508-1 and IEC 61508-4. This first part gives an overview of those principles and lifecycle processes, together with some transport-orientated guidance, illuminated by applying it to a simple, hypothetical example of the assessment of a proposed means of enabling pedestrians to cross a busy road safely.

The scope of this exercise was limited to the seven lifecycle phases relating to the specification of safety requirements but, in so doing, showed that (subject to repeating it for more-challenging rail and ATM applications) it is not only possible to apply IEC 61508 to the transport sector but it also has the benefit of ensuring a correct balance in the approach to functional-safety assessment than might otherwise be the case.

The fundamental message at this stage is that the whole safety-assessment process must start with understanding the IEC 61508 concept of an EUC, and the hazards that it presents to its environment, *before* we can go on to specify the functional and integrity required of the Safety Functions, whose role is to mitigate the EUC hazards so as to achieve a tolerable level of risk overall.

As Fowler (2015) previously showed, not understanding this message can lead to safety-assessment practices that were wholly inadequate and, as recently as April 2021, the UK Transport Secretary (Grant Shapps) tweeted the following (Shapps 2021):

> *"85% of road accidents involve some element of human error. Today we're taking the first step towards self-driving* [sic] *cars on our roads making journeys safer ..."*

Whether it was intended as a populist, or serious scientific, statement, the safety claim in the tweet is muddled thinking based on a serious *non sequitur*, as is the following related quote from the CEO of the Society of Motor Manufacturers and Traders, Mike Hawes (DfT 2021):

> *"Automated driving systems could prevent 47,000 serious accidents and save 3,900 lives over the next decade through their ability to reduce the single largest cause of road accidents — human error".*

According to UK Government statistics for the year ending June 2019, i.e. pre-COVID, there were "*1,752 reported road deaths, similar to the level seen since 2012*" (DfT 2020). To a first approximation, we could argue that, since about 1,450 of these deaths were due (at least in part) to human error, then replacing the human with more reliable automated driving systems would reduce the number of deaths caused by human drivers. However, the above two claims of an overall safety improvement would be true only if they took account of the countless millions of opportunities for fatal road accidents, i.e. the EUC Risk, which the skills and experience of the vast majority of human drivers have been amazingly successful in preventing. It follows, from the IEC 61508 principles and processes described herein, that the big challenge for the motor industry is to be able to show (with appropriate confidence and under all normal, abnormal and failure conditions) that the accident-*prevention* properties of automated driving systems (i.e. their safety functionality and performance) are also at least as good as those of the human drivers (part of the EUC control system) that they seek to replace — otherwise a huge increase in the number of road accidents might result.

Parts 2 and 3 of the series will consider the application of these ideas to ATM and rail systems.

**Acknowledgments**

The author wishes to acknowledge the considerable help, support and understanding of IEC 61508 provided by his long-standing colleague Ronald Pierce, without which this paper would not have come to fruition.



**References**

CENELEC. (1999). *Railway applications — the specification and demonstration of reliability, availability, maintainability, and safety (RAMS), Part 1: Basic requirements and generic process*, EN 50126-1, European Committee for Electrotechnical Standardization. Brussels.

DfT. (2020). *National Statistics — Reported road casualties Great Britain, annual report: 2019*. UK Government Department for Transport. Available at https://www.gov.uk/government/statistics/reported-road-casualties-great-britain-annual-report-2019. Accessed 19th June 2022.

DfT. (2021). *Government paves the way for self-driving vehicles on UK roads.* Department for Transport. Available at https://www.gov.uk/government/news/government-paves-the-way-for-self-driving-vehicles-on-uk-roads. Accessed 19th June 2022.

EUROCONTROL. (2015). *Safety Assessment Methodology, Version 2.2*. Available at https://www.eurocontrol.int/tool/safety-assessment-methodology. Accessed 19th June 2022.

European Commission. (2011). *Regulation (EU) No. 1035/2011 Laying Down Common Requirements for Provision of Air Navigation Services...* Available at: https://eur-lex.europa.eu/legal-content/EN/TXT/PDF/?uri=CELEX:32011R1035&from=EN. Accessed 19th June 2022.

Fowler D. (2015). *Functional Safety by Design – Magic or Logic?* In Proceedings of the 23rd Safety-Critical Systems Symposium, Bristol, UK. Available at https://scsc.uk/r129/7:1. Accessed 19th June 2022.

Fowler D, Pierce R. H. (2012). *A Safety Engineering Perspective*. In: Cogan B (editor) *Systems Engineering: Practice and Theory*. IntechOpen. London.

HSE. (2018). *Appraisal values or unit costs*. Health & Safety Executive. Available at: https://www.hse.gov.uk/economics/eauappraisal.htm. Accessed 19th June 2022.

HSE. (2021). *ALARP at a glance*. Health & Safety Executive. Available at: https://www.hse.gov.uk/managing/theory/alarpglance.htm. Accessed 19th June 2022.

IEC. (2010). *Functional Safety of Electrical/electronic/programmable electronic Safety-related Systems*, IEC 61508, V 2.0. International Electrotechnical Commission. Geneva.

Pierce R, Fowler D. (2010). *Applying IEC 61508 to Air Traffic Management*. In: Dale C, Anderson T (editors) *Making Systems Safer, Proceedings of the Eighteenth Safety-*

*Critical Systems Symposium, Bristol, UK, 9-11th February 2010.* Springer-Verlag, London.

Shapps G. (2021, April 28). *Today we're taking the first steps towards self-driving cars on our roads.* Twitter. https://twitter.com/grantshapps/status/1387371484423786497. Accessed 19th June 2022.

Sparx Systems. (2020). *Message Examples.* Sparx Systems Pty Ltd. Available at https://sparxsystems.com/enterprise_architect_user_guide/15.2/model_domains/message_examples.html. Accessed 19th June 2022.

# Appendix A. IEC 61508 Safety Definitions

## A.1 General Definitions

The following are some well-understood, basic safety definitions, as set out in Part 4 of IEC 61508, "IEC 61508-4" (the notes from the standard are included); in each case, the normal English usage of the words used is intended to apply, unless stated otherwise:

1. "***Harm*** – *[death], physical injury or damage to the health of people or damage to property or the environment*";
2. "***Hazard*** — *potential source of Harm;*

   *Note: the term hazard includes danger to persons arising within a short time scale (for example, fire and explosion) and also those that have a long-term effect on a person's health (for example, release of a toxic substance)*";
3. "***Harmful Event*** — *an occurrence in which a hazardous situation or hazardous event results in harm;*

   *Note: harmful events can be considered to include accidents, although the latter are usually understood to be unintentional*";
4. "***Risk*** — *the combination of the probability of the occurrence of Harm and the severity of that Harm*";
5. "***Tolerable Risk*** — *risk that is accepted, in a given context, based on the current values of society*"; and
6. "***Safety*** — *freedom from unacceptable [i.e. intolerable] risk*".

## A.2 Functional Safety

In order to fully understand the notion of functional safety in the context of IEC 61508, we next need to deal with its (possibly off-putting[20]) definitions related to the so-called equipment under control (EUC) and introduce the concept of safety-related systems (again, the notes from the standard are included):

1. "***EUC*** — *equipment, machinery, apparatus or plant used for manufacturing, process, transportation, medical or other activities*";
2. "***Environment*** — *all relevant parameters that can affect the achievement of functional safety in the specific application under consideration and in any safety lifecycle phase;*

   *Note: this would include, for example, physical environment, operating environment, legal environment, and maintenance environment*".
3. "***EUC Control System*** — *system that responds to input signals from the* [EUC] *... and/or from an operator* [and /or from the EUC's operational environment[21]]

[20] We will show that a broader, system view of that which is "under control" can be helpful in understanding how the essential principles of IEC 61508 can readily be applied to a diverse range of safety-related sectors.

[21] The clause in brackets is not included in the IEC 61508 definition - we have added it for completeness, as part of the broader, system view of that which is "under control".

*and generates output signals causing the EUC to operate in the desired manner*";

4. "***EUC Risk*** *— risk arising from the EUC or its interaction with the EUC Control System;*

*Note 1 The risk in this context is that associated with the specific harmful event in which E/E/PE safety-related systems and other risk reduction measures are to be used to provide the necessary risk reduction, (i.e. the risk associated with functional safety).*

*Note 2 The EUC Risk is indicated in Figure A.1 of IEC 61508-5. The main purpose of determining the EUC Risk is to establish a reference point for the risk without taking into account E/E/PE safety-related systems and other risk reduction measures*".

5. "***Functional Safety*** — [that] *part of the overall safety relating to the EUC and the EUC control system that depends on the correct functioning of the safety-related systems and other risk-reduction measures*".
6. "***Necessary Risk Reduction*** *— risk reduction to be achieved by the ... safety-related systems and/or other risk reduction measures in order to ensure that the tolerable risk is not exceeded*";
7. "***Overall Safety Function*** *— means of achieving or maintaining a safe state for the EUC, in respect of a specific hazardous event*";
8. "***Residual Risk*** *— risk remaining after protective measures have been taken*";
9. "***Safety Function*** *— function to be implemented by a ... safety-related system (qv)* [and/or] *other risk-reduction measures, which is intended to achieve, or maintain, a safe state for the EUC, in respect of a specific hazardous event*";
10. "***Safety-related System*** *— designated system that both:*

*- implements the required safety functions necessary to achieve or maintain a safe state for the EUC; and*

*- is intended to achieve, on its own or with other safety-related systems and 'other risk-reduction measures', the necessary safety integrity for the required safety functions;*

*Note 1: The term refers to those systems, designated as safety-related systems, which are intended to achieve, together with the other risk reduction measures, the 'necessary risk reduction' (qv) in order to meet the required 'tolerable risk' (qv).*

*Note 2: Safety-related systems are designed to prevent the EUC from going into a dangerous state by taking appropriate action on detection of a condition which may lead to a hazardous event. The failure of a safety-related system would be included in the events leading to the determined hazard or hazards. Although there may be other systems having safety functions, it is the safety-related systems that have been designated to achieve, in their own right, the required tolerable risk ...*

*Note 3: Safety-related systems may be an integral part of the EUC Control System or may interface [directly] with the EUC .... That is, the required 'safety integrity level' (qv) may be achieved by implementing the safety functions in the EUC Control System .... or the safety functions may be implemented by separate and independent systems dedicated to safety.*

*Note 4: A safety-related system may:*

*a) be designed to prevent the hazardous event (i.e. if the safety-related systems perform their safety functions then no harmful event arises); and/or*

*b) be designed to mitigate the effects of the harmful event, thereby reducing the risk by reducing the consequences* [of that event]*;*

*Note 5: Safety-related systems can be divided broadly into safety-related control systems and safety-related protection systems*"[22].

11. "***Safety Integrity*** *— probability of a ... safety-related system satisfactorily performing the specified safety functions under all the stated conditions, within a stated period of time*".
12. "***Safety Integrity Level*** *— discrete level (one out of a possible four), corresponding to a range of safety integrity values, where safety integrity level 4 has the highest level of safety integrity and safety integrity level 1 has the lowest;*

*Note 1 The target failure measures (see 3.5.17) for the four safety integrity levels are specified in Tables 2 and 3 of IEC 61508-1.*

*Note 2 Safety integrity levels are used for specifying the safety integrity requirements of the safety functions to be allocated to the E/E/PE safety-related systems.*

*Note 3 A safety integrity level (SIL) is not a property of a system, subsystem, element, or component. The correct interpretation of the phrase "SIL n safety-related system" (where n is 1, 2, 3 or 4) is that the system is potentially ... capable of supporting safety functions with a safety integrity level up to n*".

13. "***Target Failure Measure*** *— target probability of dangerous mode failures to be achieved in respect of the safety integrity requirements, specified in terms of either:*

*– the average probability of a dangerous failure of the safety function on demand, (for a low demand mode of operation);*

*– the average frequency of a dangerous failure [h-1] (for a high demand mode of operation or a continuous mode of operation)*".

[22] In general, the former can be considered to maintain a continuous or continual safe state for the EUC whilst allowing the EUC to continue to execute its normal functions whilst the latter, when required, puts the EUC into a *safer* state even if this interrupts the normal functioning of the EUC

ISSN 2754-1118 (Online) — ISSN 2753-6599 (Print)

# The Layered Enterprise Data Safety Model (LEDSM)

## A Framework for Assuring Safety-critical Communications

**Nicholas Hales**

Retired C.Eng MIET

## Abstract

*Lt. Kermit Tyler was warned of the approach of a large flight of aircraft toward Pearl Harbour. The radar operators were tracking Japanese planes coming to attack the base, but the operator failed to make clear the size of the formation and Tyler did not pass on an alarm of "attack imminent". In the case of the "9/11" attack on New York's Twin Towers, the intelligence agencies did not share relevant information. These problems, and many more like them, are caused by a lack of planning of the safety-related communication network in advance. There is a need to plan horizontal protocols to communicate with other organisations and vertical protocols to communicate effectively within organisations. The Layered Enterprise Data Safety Model (LEDSM) is a way to develop safer networks of communication of any type, verbal, telephone, internet, etc., or a mix of types, so that the risks of failures to communicate are considerably reduced. The initial idea is taken from the Open Systems Interface. This paper takes the reader from initial concepts through ten sections of increasing learning. These sections show how, even with increasing complexity, the principles involved provide increased confidence that risks are minimised. Worked examples are provided to increase insights into the numerous possible applications.*

## 1 Introduction

The ad-hoc development of communication networks involving people and systems has proven to be a contributory cause of many accidents. The Chernobyl meltdown, (Higginbotham 2019), the terrorist bombing of the Ariana Grande concert in Manchester (Collins 2021), and the mid-air collision over Überlingen, (2002 Überlingen mid-air collision 2022), all dealt with in more detail later in this paper, are examples of tragedies that need not have happened, and the risks would have been considerably reduced if communication protocols, such as LEDSM requires, had been in place.

On each occasion improvements are made to prevent such a thing happening again, but up until now, no formal method of thinking about mixed human and system safety-critical networks involving communication, that begins at initiation of an Enterprise and ends when the Enterprise ends, has been proposed. STAMP (System-Theoretic Accident Model and Processes) and STPA (System-Theoretic Process Analysis) methods are improving analysis, according to many studies, but have not linked to post-production communication. The most surprising 'Black Swan' is perhaps the one that people see and admit exists, but then forget about later to such an extent that, when it occurs again it appears to be a 'Black Swan' again. It is the one that exposes the poor communication planning of an organisation, be it engineering, political, emergency service, or medical related. It shows little was learned from the first experience. It is easy to believe that all the communication channels you will need in the event of an incident or emergency have been put in place. But when the emergency for which the communication tools were acquired comes along it is sadly all too often shown to be the case that the planned communication fails for one reason or another and, like the discovery of black swans in Australia, everyone is surprised.[23]

It is common for people to think that they know what they are doing when they establish their contacts and methods of communication for projects, events, and products that are to go into service. The facts show that time and again, disastrous consequences are the result of this over-confidence in what appears to be simple planning. Usually, the necessary communication has many nuances that are easily missed, so a formal approach is likely to improve things. It is easy to think and be seduced by statements such as, "Well, we will all have mobile 'phones to keep in touch with each other".

The Layered Enterprise Data Safety Model (LEDSM) method for developing communication networks for safety-critical networks is designed to add a level of formalism to the design process without becoming too technical, recognising that most people are not safety-criticality professionals. Despite the excellent technical progress of computers and telephony over the past thirty years, the use of communications has failed those dependent on its successful application. It is scandalous that so much can be provided by engineers to fulfil society's eagerness for communication systems, while in critical applications, behaviours required to be understood, in order to operate those communication systems effectively, are not taught well enough, usage is not planned, and development is not maintained. To fill those needs, LEDSM represents a useful safety communications management method to add to the toolbox of safety engineering.

Notable other work in this field for emergency communications includes the following:

- "Communication challenges in emergency response" (Manoj and Baker 2007).
- "A systematic approach to improve communication for emergency response" (Dilmaghani and Rao 2009).
- "Challenges of emergency communication network for disaster response" (Huang and Lien 2012).
- "Wireless technologies for emergency response: A comprehensive review and some guidelines" (Pervez et al. 2018).
- "Data-Centric Safety: Challenges, Approaches, and Incident Investigation" (Faulkner and Nicholson 2020).

Any improvement to current approaches needs to facilitate identification of potential deficiencies in communications, and so allow their contribution to risk to be reduced. This is where Why/What Because Therefore Reasoning, (WBTR) is useful, (introduced in the next Section). Then the need is to manage the communications system over time; monitoring, assessing and triggering change when appropriate, which LEDSM facilitates, iterating back to WBTR when deficiencies are apparent.

[23] In this paper, the central concept is the use, in communication involving humans, of an already existing model for electronic communication, the Open Systems Interconnection, (OSI), model redefined as LEDSM. The term 'protocol' appears alongside the descriptions of the LEDSM throughout this paper and the term is used to describe analogous arrangements to those best described by using the Cambridge University Dictionary definition "*a computer language allowing computers that are connected to each other to communicate*". Protocols in this paper means, "language formally written down and agreed between persons involved in the network allowing them to communicate safely regardless of the media used to carry the message."

## 2 Why We Must Improve Communication Networks

In the modern world we rely on networks. Business networks, networking with other people, and networks related to telephony and electronic communications. All are there to help us work better but, in situations in which critical problems can arise, risks of not receiving important data need to be treated differently. LEDSM will help in identifying important data, important people involved in providing you with information, and the machines that are critical to transmitting that data to and from you, and to and from others.

A simple example of where LEDSM is useful is when there is a pressing need for newly-communicating Enterprises to pass important data. What happens, for instance, when the communications are almost guaranteed to be incompatible — for instance, when new allies cooperate in a battlefield situation. In the event of incompatible communication systems, allies would almost certainly resort to verbal communication, and this is precisely where LEDSM can be helpful. The development of new systems to make the previously incompatible systems of the allies compatible could take years to develop, whereas LEDSM would only take weeks to implement, and could be done to a sufficient standard for some confidence within days.

The large number of incidents indicates that safety-related communication is not treated with sufficient care by many institutions. By formalising the communication expected, using LEDSM, as proposed here, greater focus and care may be taken. If planning of communication is done more formally, within organisations and in communicating with other organisations, and processes adhered to in practice, many deaths and severe injuries can be avoided.

There are two distinct ways to deal with safety-related information delivered by networks of systems and people. One can prepare for incidents, in the belief that on the day when an incident occurs, it will be a relatively simple affair to receive information from those on the ground and hand out orders to them in return, because "one knows one's job." The other way is to realise how important delivering the right data at the right time is and acting on that understanding. The abstract of the 2019 Data Safety Guidance, produced by the Safety-Critical Systems Club (SCSC 2019), states "*Data, as distinct from software and hardware, has been a contributing factor in many accidents and incidents. The impact of data-related issues continues to grow as we rely more and more on data-driven systems.*" Primarily, this paper is about looking at how humans may better process information within data-driven systems. It does not, however, exclude machines and electronic systems involved in communication. It may therefore be of some use to those developing such systems in which humans are not involved, especially if human-like decisions have to be made and communicated. That should make for safer environments for all.

The following sections describe how the LEDSM has adopted and mirrored the Open Systems Interconnection (OSI) model of the electronics world (ISO/IEC 1994). It should improve both the way we utilise communication systems, and the way we interface with communicating systems when designing essential safety-critical communication networks at many levels of granularity, providing greater confidence to both management at more abstract levels and those more deeply involved in delivering safe processes.

LEDSM communication can be seen as sets of protocols within Enterprises and between Enterprises. In addition, it can be used in conjunction with a method of identifying potential problems in communication, through a form of disciplined brainstorming. Built on Why-Because-Analysis (WBA), used to analyse accidents (Causalis 2018), that method is WBTR, mentioned at the end of the Introduction above. It takes an *a priori* approach, as opposed to WBA's post-accident analysis, that might be considered to be akin to the well-established HAZOPS (Hazard and Operability Study) used in the chemical industry. In this analogy, data is like a fluid flowing between processors or processes. Section 10 gives a small worked example of WBTR.

It is important to let everyone state what information they expect, and what information they can guarantee they will deliver, when considering safety-related issues. If what you want in terms of information or data, from any individual, either in your Enterprise or in another Enterprise, is not in the data that any other individual or Enterprise is declaring they can guarantee, then more analysis needs to be done. Solutions thus derived from that analysis need to be implemented in the form of Dependency-Guarantee Relationships, (DGRs). A DGR involves person or system 'A' guaranteeing to another person or system 'B' that the data they are dependent upon will be provided in a timely and accurate manner, i.e. within a certain time frame and reliably enough to act on. DGRs are described more fully in the context of LEDSM in Sub-section 3.8.

Machine to human, and vice versa, communication is notorious for errors. If the tasks are simple then machines can make all the critical decisions, even using Bayesian probabilities to make decisions when an intelligent knowledge base system is included within the decision-making loop. However, if the situation is so critical that it requires a human intervention, known as Human-in-the-loop (HITL) systems, then the decision over what is said or done is inevitably subjective, to a greater or lesser extent, and therefore likely to introduce errors, especially in stressful situations. Furthermore, if the machine accepting human intervention then has to pass information, via communication links, to another machine, that will subsequently provide that message to another HITL system, then the possibilities for errors multiply. Hence it is vital that agreed protocols between organisations in any particular loop, i.e. about what safety-related data will be passed within what time frame, are established. Only by knowing what cannot be correct, or is questionable, will a data receiver be able to take appropriate actions, including asking for a repeat sending, or clarification, if necessary. Systems risks can soon increase when a HITL is essential, so one should always permit an independent view of any proposed emergency procedures and processes.

The importance of safe communication can be illustrated briefly here. As already mentioned, there is a technique used to analyse accidents called WBA. WBA has been used to analyse the incident when 193 people died due to the Roll-On-Roll-Off, (RoRo) ferry, Herald of Free Enterprise, sinking after leaving harbour with its bow doors open (Why–because analysis: Example 2006). The findings tie closely with the court of inquiry conclusions except in one respect, that the official inquiry identified "*a general culture of poor communication in Townsend Thoresen*" (MS Herald of Free Enterprise 2022). It is this conclusion that shows how important the planning of good communication must be, and LEDSM assists in formalising this process. Once a network has been designed, and before development, it should be examined using the WBTR technique, improvements made if necessary, then the communication paths documented using LEDSM and the necessary agreements, or protocols, formalised. One of the most useful aspects of LEDSM is that it is flexible and new communication paths are easily added in. If anything is missed, a later review before implementation may show other paths are necessary, but bear in mind that this is expensive in system development, though relatively cheap in the case of exclusively human communication, by re-training and creating more channels. A review of the original WBTR should be performed when additions to or subtractions from networks are made, looking at each new possible failure and checking that the response is appropriate and, if not, continuing with developments.

Identification of risks can be a highly technical process and may require specialist knowledge and/or experience. In such cases, the analysis should not be undertaken by non-professionals. However, just from our casual lives many of us are familiar with safe engineering principles and these are embodied and evident in the use of LEDSM, making it useful for non-professionals as well. Implementing networks of people and systems employing LEDSM will reduce risks even if developed by those unfamiliar with safety engineering, but employment of professionals to assist and review would be wise in any event to ensure all aspects are considered.

To summarise, this paper is intended to instruct on how to reduce the risks involved in network communication by exploration of examples of the use of LEDSM and associated techniques. It achieves understanding through taking the reader through a gradual development of learning. On finishing reading the paper, it should be easier to integrate the use of WBTR, LEDSM and DGRs to create less-risky data and information transmission systems, even when they include humans who often are involved in transmitting information.

# 3 Management of The Layered Enterprise Data Safety Model

## 3.1 Overview

This section describes how LEDSM both within an Enterprise and with other Enterprises, should be managed. There are two viewpoints that must be understood:

- The first is the overall management of an Enterprise's LEDSM, and what Senior Managers must know to be effective in managing their Enterprise's emergency communications and consequent actions.
- The second is management at lower levels of the Enterprise where the vast majority of interaction with other layers within an Enterprise and with other Enterprises takes place.

Reading this section will help management to understand the principles they need their staff to operate by, but also assist technical staff and safety behaviour practitioners in understanding why management will require adherence to strict codes. This continues previous work published in the Safety Critical Systems Club's Newsletter (Hales 2020) and the Proceedings of the Safety-critical Systems Symposium (Hales 2021).

## 3.2 The Reason the Term 'Enterprise' is Used Instead of 'Organisation'

The primary reason the term Enterprise was coined in the development of this safer data communication management technique is that it neatly parallels business. In business, 'enterprises' can be anything from huge corporations down to partnerships of two people or even just sole traders, such as market-stall holders, whereas 'organisation' is not a term that could be used to describe sole traders.

For the convenience of design and realisation of the essentials of safe communication in any system or system of systems, some individuals will be masters of their own domain and will know with whom they need to communicate in other domains. They may be Enterprises by themselves, as will a person appointed to lead a project before any design work starts. Examples of that are a field researcher hunting down the facts of a viral outbreak, which is briefly examined later in this paper, or the manager of the Chernobyl Nuclear Power Plant project when it was first proposed, Director Victor Brukhanov.

### 3.3 The Layers

Any Enterprise may have up to seven layers. This is a somewhat arbitrary figure, but the layers used in the OSI for communicating electronic systems, are seven, so it is considered that should be sufficient at a maximum, logically, as a model on which to base safe human and human-system communication. An Enterprise though could just have one layer, in a similar fashion to how an Enterprise can be a sole trader. On the other hand, the seven-layer limit helps to restrict bureaucracy, an unwanted characteristic in safety-related communication, in very large organisations.

### 3.4 Protocols

Inherent within LEDSM is the use of Protocols. These govern communication between the (up to) seven layers of an organisation with safety-related communication requirements. They also govern communication between identical layers in other organisations. For example, there will be a protocol between the management of each Enterprise and the management of every other Enterprise with which direct communication takes place. The protocol at Management level will indicate to another Enterprise, at an abstracted level, what information it will be expected to supply and what is expects to receive. That will also cover what the lower layers will expect to receive and deliver, though of course the lower layer protocols will have more detail.

As an example (see Figure 1), a Nuclear Power Enterprise management protocol may state that information as to whether an emergency evacuation of a building was underway may be sent at Level 2 from the nuclear facility, Enterprise 1 in the diagram, to the emergency services Enterprises, (police, fire, military and ambulance), Enterprise 2 in the diagram, (only one representative emergency service shown for clarity). The colours used are standardised for diagrams in this paper, yellow for management, blue for supervisory and beige for action; note, where condensed and when one person is the Enterprise, the highest layer colour is used).

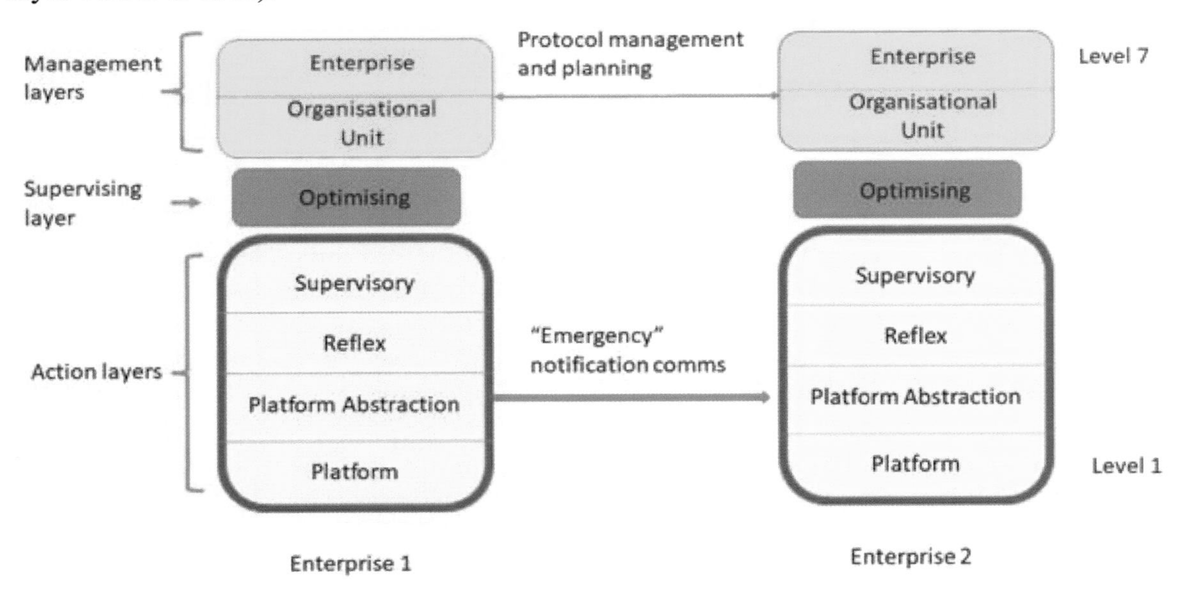


**Figure 1 ~ Illustrating the 7 Layers and Communication Expected**

An explanation of the terminology used in defining the layers is given in Sub-section 3.10, (the top three layers) and Section 5 (all 7 layers with OSI equivalents). For the moment they are for clarity and later reference only.

Those emergency services Enterprises will agree in the protocol to provide emergency services with a given timescale and agree to notify the nuclear facility of any delays. The Management level of each protocol will also have protocols, cascading down their own Enterprise to Layer 2, of what information should be supplied to keep management informed of progress. Ultimately, management oversees and approves the form of the protocols to all levels. However once those are in place, management should only be actively involved in communicating with management in other Enterprises and Layer 6 in their own Enterprise, the level immediately below Management, both (usually in slow time) to adjust protocols as processes and technology evolve.

The rest of the emergency response will be governed by protocols already in place and should involve automatic or well-rehearsed responses. This reflects how much of what is done in computers to ensure accurate communication is done without the user at the top level having to know what is happening. Figure 2 illustrates the protocols as applying to just one (any) layer and its adjacent layers.

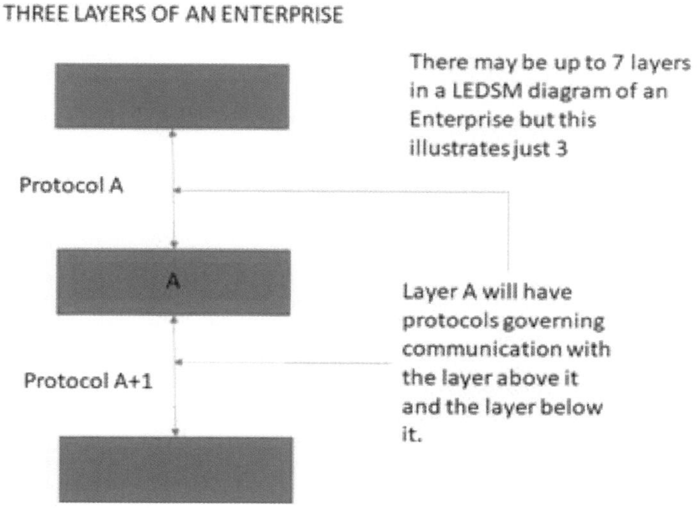

THREE LAYERS OF AN ENTERPRISE

There may be up to 7 layers in a LEDSM diagram of an Enterprise but this illustrates just 3

Protocol A

A

Layer A will have protocols governing communication with the layer above it and the layer below it.

Protocol A+1

**Figure 2 ~ An Illustration of the Position of Protocols within an Enterprise**

In Figure 2, the box labelled A represents a layer of LEDSM within an Enterprise and it will have two vertical protocols, one for sending and receiving data from a lower level, (more toward the instantaneous emergency response level), and one for sending and receiving data to a higher level (more toward the management level). The protocol would describe in what circumstances instant decisions can be made and communicated outward from the Enterprise, via the lower level if necessary. Note that no horizontal connections to other Enterprises are shown here, so it may be thought of as being a snapshot of an Enterprise under development.

The interface between vertical and horizontal layers in an Enterprise must be governed by the agreed DGRs. The DGR is dealt with later in Sub-section 3.8.

### 3.5 Dealing with Large Volumes Supplied by Some Communicating Enterprises

A very large and reasonably effective communicating organisation with many thousands of participants receiving notifications and e-mails, can be viewed as one Enterprise when being linked into a network model. An example of such is the Programme for Monitoring

Emerging Diseases (ProMED) e-mail that often provides *ad hoc* information for those involved in viral outbreak prevention. This is preferable to trying to make an absurdly tangled wiring diagram of who is likely to want data from whom. This is illustrated in Figure 3.

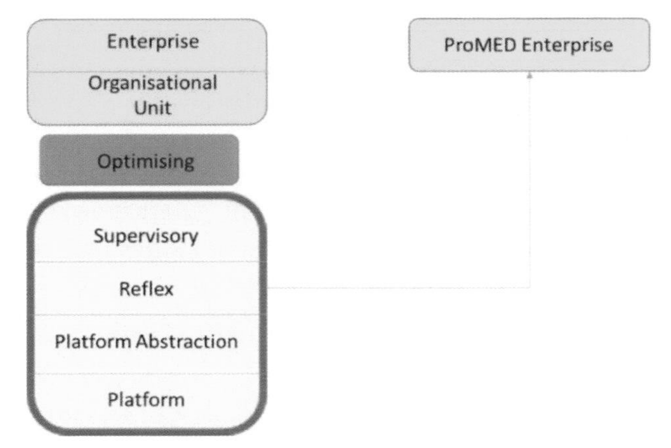


**Figure 3 ~ Linking to an Enterprise with Multiple Connections**

### 3.6 Keeping the Network Visualisation to the Minimal

Off-line and for their own peace of mind, researchers, politicians, engineers, scientists, medics and commercial businesses can generate maps of their networks at levels of granularity that are different from those handed down by Protocols within their Enterprise. LEDSM thereby enables them to see how, with some thought applied, they can propose a case, for instance, to the higher levels vertically above them in their Enterprise for further communication links to be formally established with other organisations through horizontal Protocols. Management therefore is encouraged to be open to suggested improvements to Protocols. A diagram of the connections any individual has can be readily seen using LEDSM Enterprise diagrams and held by the individual without needing to know non-essential links across the network that others in their enterprise have. Non-essential in this context means 'no need to know' though, of course, as we are developing and mapping out a safety-critical communication network, all connections should be essential. It is simply that while access to the greater network may be necessary for one's understanding, a downloaded diagram of one's own connections will be kept as simple as possible so that comprehension of the completeness of one's needs are not stifled by the complexity of the network.

An example of this is when a single person operates a safety-critical data and information communication centre, such as a field worker for a satellite office, as in Figure 4 in which a World Health Organisation, (WHO), worker operates out of Jakarta to track and identify virus outbreaks. This scenario was covered in greater detail during the Twenty-ninth Safety-Critical Systems Symposium (Hales 2021). Note, of course, that the WHO worker acts as the equivalent horizontal protocol owner for all roles in other Enterprises where the layers have been allocated to different individuals. This is because the worker may have to have different protocols to communicate with an Enterprise, depending on the data to transmit. In this example the Field Worker, as a sole individual in the Enterprise, may ordinarily simply communicate with say Level 3 in another Enterprise, but when conclusions about a virus outbreak are reached, the worker may have protocols to deliver a message to the Level 7 management. An analogy would be for a sole trader in a market, talking to a high-level member of sales team to order clothes to sell, but when not receiving needed stock at an agreed time may wish to communicate with the management of the supplier.

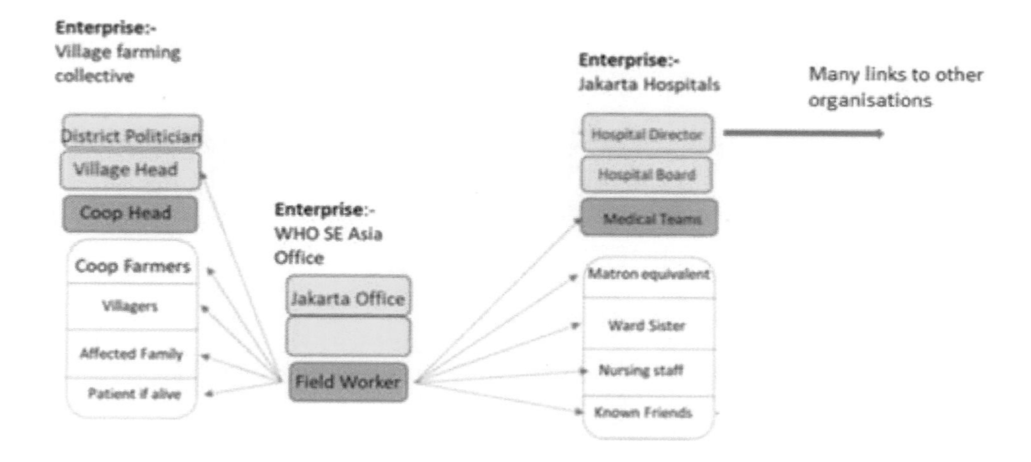


**Figure 4 ~ An Illustration of a Single Person's Connections**

Connections can be expanded or reduced by the use of LEDSM alone or LEDSM and WBTR, which was briefly described earlier and is set out in more detail in Section 9. Managers at the top level of Enterprises using LEDSM should require WBTR be carried out at initiation of the network and at intervals when changes are made to help reduce the risks involved in communication channels to prevent accidents.

### 3.7 Levels Senior Management need to Understand

Management need only look at the top level of other Enterprises and at the layer below their own. In effect, senior management can condense any image of a LEDSM structured communication system to the bottom and top layers as illustrated in Figure 5 for a full seven-layer model.

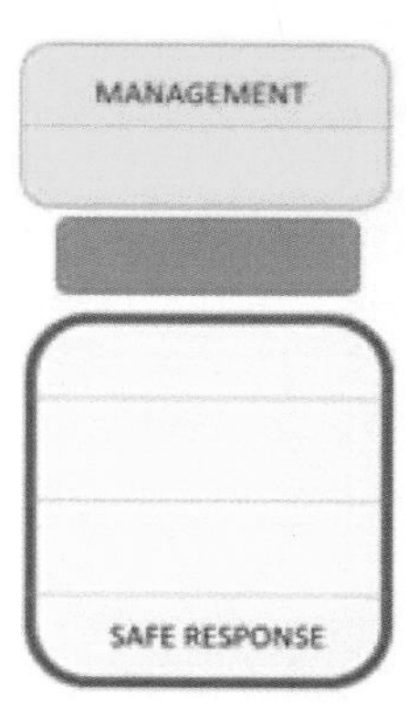


**Figure 5 ~ A Simplified Management View of Enterprise Safe Communications**

In electronic communication systems governed by the OSI, this is equivalent to what is to be transmitted, known at the user level (in this case management), and what is actually transmitted down wires to another communicating system, at the internet or intranet level (in this case other systems and people). All the protocols between the two are there to facilitate the effective practising of safety-related actions as required by management.

Protocols within the Enterprise dictate the communication and action necessary to achieve the management goal at the lowest 'Safe Response' level, which should be guaranteed to other Enterprises. Horizontal communication protocols will exist between one or more layers when communication to other Enterprises is involved.

## 3.8 Dependency-Guarantee Relationships

In safety-critical systems, the DGR is a formal and strict relationship in which one part of, for instance, a software system driven using two hardware processors, "guarantees" to another part that the software it runs will deliver a value within a given time constraint. That other part wants that guarantee because it in turn "depends" on that value being delivered, within a given timescale and to a degree of accuracy and correctness, in order to fulfil its own obligations to the system.

A DGR means that, as the safety-related function of one layer is dependent on information or data from another layer, the layer delivering data must offer a guarantee that such data or information will be made available, and it will be timely and safe. Each layer requiring data must state precisely what its needs are so that its 'Dependency' is understood by delivering layers. Protocols can then be developed. LEDSM uses the OSI model as the model for protocols that govern the relationship between Management objectives and the delivered information passed to other Enterprises, because in electronics each layer can only deal with data in the format it has been designed to accept. This applies to safety-related data communication systems equally. Management must be able to have absolute confidence that best possible responses are made, and the protocols of LEDSM provide the means to achieve that. Hence the term 'protocols that exemplify the principle of DGRs' is a useful way to review communication preparations for an emergency within an Enterprise and with other Enterprises.

Each layer is governed by these DGRs in a vertical sense in their own Enterprise and the Dependency and Guarantee statements should be iterated until an agreement is reached, known as the protocol, between it and adjacent layers. Difficulties that managements of Enterprises have in formulating policies with other Enterprises may create a need to modify the DGRs between layers, thus modifying the Protocol. In most situations that are safety-critical, it is preferred that Enterprise leaders are flexible enough to see the sense in minimising any difference between the needs of layers of another Enterprise and what their Enterprise actually provides, after negotiations are completed. What they do provide should be the maximum possible of the stated requirement of the other Enterprise.

The degree to which one relies on the declaration of guarantees of delivery of information need to be exposed to scrutiny, as does one's own needs, preferably by an expert third party, to ensure one has not missed issues due to the effect that familiarity tends to blind us to our own faults when things get complex. This blindness can be a result of experience making one assume that technological improvements can only enhance, and not destroy, one's carefully developed processes.

## 3.9 Creating the Initial Protocols by Adapting from Other Sources

Here we will examine already published standards and protocols on behaviour to see how, rather than reinvent the wheel, we can adapt statements to the relatively new field of Data and Information Safety Management. When implementing a LEDSM for their organisation, those involved may refer to this section to provide example wording to support activating appropriate protocols in their Enterprise.

As a brief example, many of the stated principles of professional engineering organisations fit well with LEDSM principles. For instance, the Engineering Council, on the issue of the environment states, "*Seek multiple views to solve sustainability challenges*" (Engineering Council 2021). Equally, seeking many opinions is the cornerstone of the recommendation to constantly review the LEDSM network one establishes and the brainstorming that involves WBTR. The visual nature of LEDSM should simplify doing that. So, the principle to adopt is to "Seek multiple views to solve data and information supply and delivery challenges". Another Engineering Council principle, slightly revised to apply to LEDSM, would be "Manage communication of data and information to minimise any adverse impact on people and the environment".

LEDSM is designed to reduce risks from all manner of hazards and a source of statements which may be rehashed to provide equally applicable principles is the Hazard Analysis guide of UK Defence Standard DEF.STAN.00-56 (UKMoD 2007) as it was at Issue 4. For instance, Clause 6.5 in that standard may be revised to become, "The Enterprise management, together with those in the Enterprise charged with responding to accidents and emergencies, i.e. those at the lower levels of the LEDSM diagram — levels 1 to 4, shall implement measures to provide the opportunity for effective stakeholder representation during safety management activities". So data communication management becomes an obligation of management to ensure their staff have established paths to communicate with many other Enterprises that need to know what data is available about the accident or emergency preparations or occurrences. At least all stakeholders need to know they were consulted.

DEF.STAN.00-56/4 Clause 7.1 also can be revised simply to: "Enterprise management must retain evidence that data communication tasks within their control and that influence safety are carried out by individuals and organisations that are demonstrably competent to perform those tasks".

Another typical protocol applicable to link Enterprise management with engineers, medics, researchers, or emergency workers may be taken from DEF.STAN.00-56/4 Clause 8.1.1, to become, "The Data and Information Transmission Management Plan shall detail the specific actions and inter-Enterprise protocols required to operate Data and Information Safety communications both within an Enterprise and externally. It shall ensure Data and Information Safety is achieved and maintained."

Lastly, as another example, Clause 9.1 of DEF.STAN.00-56/4 can be revised to read as follows: "The protocols for safe transmission and receipt of Data and Information shall consist of a structured argument, supported by a body of evidence that provides a valid case, comprehensible by all stakeholders that an Enterprise is producing and accepting Data and Information that is safe at the time of review, recognising that changes may occur between reviews in both the task the Enterprise is seeking to achieve (for example, once a virus is conquered, another may be on the way of a totally different genetic make-up within weeks), and the environment in which the Data and Information communication tasks are to be achieved". For example, some previously active Enterprises may have disappeared, or governments may have changed, etc. This is, of course, an indication that reviews of the inter-Enterprise network should be undertaken quite regularly and, depending on the need for the networking, may be as little as every month or as much as every six months. For disease control, one month may be appropriate, for emergency services responses to disasters or terrorist incidents, six months may be appropriate.

In essence, once LEDSM is understood, policy can be created to ensure it is conducted appropriately and safe communication DGRs maintained.

## 3.10 How a LEDSM Enterprise Starts as a Management Level

It is proposed that, when the Chernobyl Nuclear Power Plant was first initiated, it would have been possible to avoid some of the catastrophic decisions that led to the disaster by using LEDSM. As with many disasters, it is the initial direction and influence of management at which an error is made, with many subsequent errors compounding the problem. Just like in safety-critical software development, if the potential for disaster is not recognised, safety issues become very expensive to address later in the development and may even be ignored until disaster strikes. This is discussed here in more detail.

Figure 6 shows some of the main characters that were influential in the chaos the led to the Number 4 Reactor meltdown at Chernobyl, but here they are embedded in an LEDSM diagram that, had it been constructed at the time, would have reduced the inherent danger of communication failures that led to the accident. It becomes obvious what went wrong, and what should have happened. Much of the rest of this sub-section is derived from "Formalising Communication on Potentially Catastrophic Safety Projects" (Hales 2020), which may be consulted for further information if required.

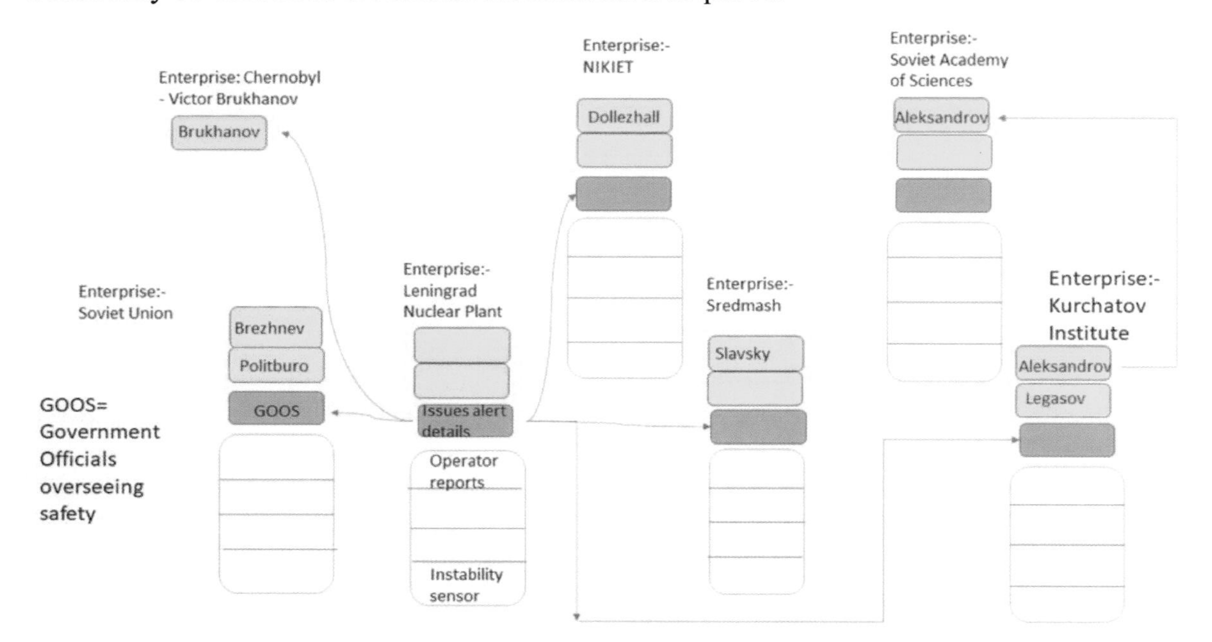


**Figure 6 ~ Chernobyl Construction-related Enterprises in 1970 and Illustrative Links**

The definitions of the higher management levels of the model and their relation to the nuclear command structure in 1980s Russia are examined next. Only the top three levels of the model are discussed. Detail of the OSI descriptions of all seven layers, and how they relate to their interpretations in LEDSM usage for any given project or system, are given in Section 5.

**Layer 7** – The 'Enterprise' layer

The Enterprise layer is the corporate entity, which in the case of the Soviet Union under the Communist Party, can be taken as the entire country, as it saw itself as having a common purpose. The layer is defined as being responsible for the planning and execution of large-scale changes to the infrastructure, responding to changes in legislation, setting and maintaining standards, procedures, and competency requirements.

In terms of monolithic Enterprises, and the Soviet Union represents such a creation, changes to the infrastructure and responding to changes in legislation can be seen as influences from outside the Enterprise, i.e. competition forcing the Soviet Union to produce cheaper electricity than Western countries in order to maintain living standards at a similar level to those other countries. Setting and maintaining standards, procedures, and competency requirements, more than educational and experience requirements at this level, implied membership of the Communist Party in those days. This, of course, meant that some talent would never achieve greatness for those who were considered less than committed to the party. The cracks in the design process, showing up later, occur because data has a source and a sink and, if the source is less than competent, the sink, whether human or machine, will not operate to best intent, though one may 'get away with it' for a while.

**Layer 6** – The 'Organisational Unit' layer

This layer is described in the OSI as being responsible for delivery of the planned service. In the Soviet world this naturally describes the Government. The layer is described as not playing any part in the day-to-day running of the system, whatever it may be. In the OSI it encodes, encrypts, and compresses data for transmission, so in the LEDSM context it may be translated as acting between higher management and safety personnel of lower levels to ensure the lower levels understand the higher-level expectation and also that management understand where the problems may lie in their proposals, as fed up through the Enterprise layers to them, by the lower-level personnel. Of course, in authoritarian societies exactly the opposite often happens; interference starts, persons lower in the hierarchy are not listened to and unfortunately disasters can easily follow. Authoritarian tends to imply an understanding of the Enterprise layer above, but very little hands-on understanding of the problems with which those delivering these ambitions, in the lower layers of the system, have to deal, since everyone is behoven to those above in such a way that argument is suppressed.

This layer is likely to become involved in the short-term operation of the system in response to a serious incident that causes substantial impact on the delivery of the system. In the case of the explosion in Reactor 4 at Chernobyl, this definition precisely describes the substantial impact on the Soviet Enterprise intent, namely to create machines upon which a perfect society could rely to run smoothly and be able to outcompete the West. Hence, what was required was much closer safety data definition and competent communication. The point is that Suitably Qualified and Experienced Personnel (SQEP) are needed, not necessarily party members if the SQEP is more experienced, or more highly regarded technically, than the party member. That did not happen, which is a lesson to managers that, if understood, will reduce risks to any safety-related system or communication network in which they are involved. This applies in Western Enterprises too in that likes and dislikes of personnel should not influence the choice of who is to oversee safety.

**Layer 5** - The 'Optimising' layer

The Optimising layer is described as the most sophisticated control layer. It respects the performance and safety constraints of the underlying system and the information contingency plans. Crucially, this layer may be described by the words: "information demands on the Optimising layer are high, requiring a full understanding of the underlying system, the planned service and contingency plans." This layer, in Chernobyl terms, represents "The Nuclear Experts".

### 3.11 The Chernobyl Communication Problems

Below are some of the involved characters and **Enterprises** as defined in the original article (Hales 2020):

- Anatoly Aleksandrov — Chairman of the Soviet Academy of Sciences, responsible for nuclear technology development. — **Classed as an Enterprise** and also Director of the Kurchatov Institute — **Classed as an Enterprise**
- Nikolai Dollezhall — director of The Scientific Research and Design Institute of Energy Technology (Russian acronym NIKIET) — **Classed as an Enterprise**
- Victor Brukhanov — Director of Chernobyl plant — **Classed as an Enterprise**
- Efim Slavsky — Ministry of Medium Machine Building (the Russian acronym is 'Sredmash') — **Classed as an Enterprise**
- Leningrad Nuclear Plant (1st RBMK design) — **Classed as an Enterprise**
- Leonid Brezhnev — The Soviet Union — **Classed as an Enterprise**

So here we have one engineered Enterprise (the Leningrad Prototype Reactor), four engineering Enterprises, an engineering project barely underway (the Chernobyl Nuclear Power Plant itself), and a political Enterprise that all the others serve. There would be more later, such as the Enterprises providing components and cement for the construction of the Chernobyl reactors, but for this snapshot they are not considered. All those named could have functioned more efficiently had data and information safety been a design criterion within the Enterprises involved. This identification of involved parties illustrates how useful the term Enterprise can be, as an all-embracing term for anything from an individual to a large organisation, including governments.

Early in the development of the Chernobyl Nuclear Power Plant, when only the director had been appointed, the LEDSM diagram may have looked something like Figure 6. As can be seen, Victor Brukhanov has been appointed director but without sub-ordinate levels at this stage. All seven layers are not absolutely necessary, as roles may be combined when not in danger of overloading one person with responsibility. When an Enterprise is first mooted, there will be only one layer, the Enterprise itself, and that layer will communicate with all other interested parties until the Enterprise starts to grow and more data intensive communications begin to be necessary. All responsibility at that stage was in Victor's hands. Figure 6, then, is how a retrospective view of the Chernobyl Nuclear Power Plant construction LEDSM network, dated to early 1970, would appear.

Problems with the first RBMK, the Leningrad Reactor, needed to be reported by the operator to whomsoever had taken the role of optimising at Layer 5. To recap, that person had to understand how it should operate, the planned service and contingency plans, how the operations are to be achieved, and what the safety plans were. Management should expect a brief on whatever problem is uncovered and the fact of it should be communicated to other Enterprises that needed to know, including the Chernobyl project, as shown by the connecting line.

In summary, management should firstly have confidence that the lower levels, 1 to 4, are aware of the vertical and horizontal protocols that they are expected to follow and know their jobs well. Secondly, they should have confidence that the network is optimised and that all possible and known stakeholder Enterprises will be alerted in the event of an incident, and are aware of what is expected of them. Pre-planning using LEDSM is a key to presenting an easy-to-follow diagram.

The importance of management connectivity can be seen. The lower four levels of LEDSM are where much of the automatic responses to emerging crises, accidents, terrorist incidents, and many other life-threatening incidents occur. It should be abundantly clear that ensuring communication of intent, including the establishment of protocols vertically within Enterprises and horizontally to other Enterprises, is an important part of management activities, ensuring safe data and information are both timely and useful when incidents occur, expected or otherwise.

The primary responsibility of management in regard to LEDSM, is to ensure that those lower down the model within their Enterprise, are aware that they must establish initial communication links and develop subsequent protocols, concerning safety-critical data and information transmission, with those people or other data provision systems, such as databases and email newsletters with which they have been instructed to communicate. Though they may also suggest to their management additional persons or systems that should also be in their network.

# 4 Reasons to Use LEDSM

## 4.1 Rules Can Be Very Important

In the book by James Martin, founder of The Oxford Martin School at the University of Oxford, warning of the terrible things that may happen if the human race does not carefully plan the future, "The Meaning of the 21st Century" (Martin 2006), he writes that if humanity is to survive, "*We need to put in place rules, protocols, methodologies* [sic], *codes of behaviour ... that will enable us to cooperate on the planet and thrive*". LEDSM could be described as being any of those as there are:

- rules governing the levels;
- protocols between the layers above and below and across to equal layers in other Enterprises;
- recommended methods on how to develop an LEDSM-based network for maximum effectiveness; and
- codes of behaviour that must be adopted by those at the ends of communication paths to operate critical communication infrastructure effectively.

The LEDSM approach will help in putting in place those needs James Martin identified in critical communication situations.

## 4.2 Example Failures

Confidence in networks that have been developed for use in emergency situations is increased by the use of protocols between participants in the network. The establishment of protocols is a very important aspect of a LEDSM network. For illustration, here are two well-understood failures to communicate effectively:

1. Using the rudimentary RADAR available at the time, Lt. Kermit Tyler of the US Army Air Corps was charged with managing the monitoring of the skies around Hawaii for approaching enemy. He was not trained fully in his role and he made the assumption that the approaching Japanese Air Force, that an operator warned him about, was just a scheduled delivery of US aircraft to add to the B-17 bomber fleet. The 'attack imminent' signal was not therefore issued (Kermit Tyler 2021). This would not have happened if definitive protocols had been in place and Tyler had been trained in their use. The question arising is why would his superiors not believe protocols should be in place, protocols that dictated

what was to be done in the event of each type of radar return? Asking him what he was thinking of when he did not alert Pearl Harbour then becomes academic. The point is that the same question could be asked of the Chernobyl Power plant operator, what on earth were they thinking when they conducted the fatal test? With Protocols in place such questions will not arise as long as everyone abides by them.

2. During the Falklands war, despite intelligence briefings that identified an Exocet attack by Super Étendards as possible, HMS Sheffield had assessed the Exocet threat as overrated for the previous two days before the attack which destroyed it. Despite HMS Glasgow, on detecting the radar signal of the aircraft, immediately going to action stations and communicating the warning codeword 'Handbrake' by UHF and HF to all task force ships, HMS Sheffield still appeared not to assess the risk wisely. On the command ship, HMS Invincible, the warning was reduced to Amber instead of raising it to Red because they had no confirmation. Seven seconds after detecting the radar blip from the aircraft, the first Exocet missile was fired, in response to which HMS Glasgow fired its chaff. HMS Sheffield did not detect the attack until lookouts reported the smoke trail of the missile. The bridge officers did not call the captain to the bridge, made no call to Action Stations, made no evasive measures, and made no effort to prepare the 4.5-inch gun, the Sea Dart missiles, or order chaff to be fired. The anti-air warfare officer was called to the operations room by the principal warfare officer, arriving just before the first missile hit (HMS Sheffield (D80) 2022). Again, what were the officers on the bridge thinking?

It should be obvious that, in either case, protocols of what data was to be communicated, what systems were to have more than one channel to communicate, and what procedures were to be followed in communicating to others, were not developed or used. Incidents like these are easily identified as in need of improved communications, but 40 years after the Pearl Harbour attack, errors that cost lives were still being made in the Falklands, and they still are around the World.

To illustrate how just informal diagrams assist in comprehension of where communication is deficient, a diagram of the Pearl Harbour communication system is given here (Figure 7). Looking at the diagram, having read this paper thus far, one should be thinking, "What were the protocols between those connections?"

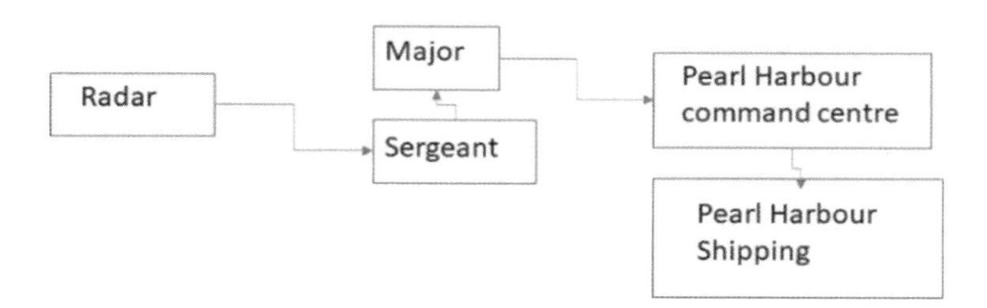


**Figure 7 ~ The Pearl Harbour Chain of Communication without Protocols**

It is apparent that with just simple protocols of what is expected of each represented participant in the chain, (except the radar which would have served its purpose most reliably by there being a reserve second radar), the attack could have been detected. These protocols would have governed what information is to be passed, to whom and within what timescales. LEDSM adds to such simple diagrams by providing more formal diagrams that are flexible and negotiable for stakeholders to contribute to. The diagram also assists in inspiring questioning during WBTR exercises.

## 4.3 Resistance to Change to More-formal Methods

Sometimes in developing networks for safety-critical issues one may encounter people and organisations unwilling to engage in the necessary protocols within and outside their Enterprises. The Peoples' Republic of China covered up the Severe Acute Respiratory Syndrome outbreak when it emerged in 2002, and it appeared to be covering up COVID-19 in the early stages of January 2021: "*By Friday 10th January it was clear that the Chinese authorities knew more than they were letting on*" (Farrar and Ahuja 2021).

The transmission of safety-related data well in advance of incidents can be mired in bureaucracy. On Saturday, April 26th 1986 at about 1am, the Chernobyl meltdown had occurred and the roof had blown off the reactor. It was not until the Tuesday following that the Soviets released any information pertaining to the disaster, as they tried to save face, yet just after 7am on April 28th, the Monday, a Swedish nuclear engineer tested his shoes at the plant he was working on, and saw they were giving readings of gamma radiation way above normal (Higginbotham 2019). This was not only a failure to deliver data at the time of the emergency; it had been known since 1983 by nuclear design experts in Moscow that the rods of the emergency shutdown briefly caused an upsurge in reactor power, but this data was suppressed and there were no protocols governing the release of safety-related information and associated lists of who should be informed. The development of LEDSM protocol would have reduced the risk of an accident.

## 4.4 LEDSM is Not Costly to Implement

Emergency responses can be poorly planned, but why is there reluctance to invest? It could be because safety is often labelled as "Costly to Implement", so there is some reluctance to delve into problems that ought to receive attention. Consequences can be deadly, as illustrated by the holding-off of emergency workers from entering the building in Manchester, England, where an Ariana Grande concert had just taken place and a terrorist suicide bomb had been detonated. The delay in responding is considered to have contributed to extra deaths. The simple reason for it was poor planning of communication. The way LEDSM helps in these situations is that, because protocols are in place horizontally between Enterprises and vertically within Enterprises, each Enterprise knows in advance whether or not the data it receives or broadcasts is complete or as complete as the protocol permits. Correct assumptions can thereby better be made if data is as yet incomplete and demands for the fuller or complete data, in accordance with the totality of the agreed protocol, progressed as communications and knowledge improve.

LEDSM is a graphic tool for the development of communication networks in disaster situations, which is designed to be understood and used, even by those unfamiliar with safety-critical systems engineering and the associated data safety issues. It reduces the likelihood of failure to communicate in a timely and effective manner in stressful situations, through better planning. The big benefit of using LEDSM, though, is that it is very low cost to implement. The return, in terms of any avoided catastrophes and major incidents, should be enormous when all the costs of a severe incident are taken into account — these include: the damaged equipment, environment or buildings; the injured or killed people (relatives of whom are likely to sue); the stressed employees (who may have to take months off work); and the costs of what can be a lengthy inquiry before a judge.

# 5 The Seven-Layer Model Description & Internal Enterprise Protocols

In this section, we look at the use of LEDSM within an Enterprise. In later sections the external protocols to other Enterprises are discussed.

## 5.1 The OSI Seven Layers Described in LEDSM Terms

**Figure 8** shows the seven layers of the LEDSM. The summary description below that is of each of the seven layers of the OSI (ISO/IEC 1994), see also Shaw (2022), for example. At the side of the description, in bold, are examples of the usage of that layer in a LEDSM-based Enterprise's safety-critical operations and communications. The name given in the LEDSM diagram to each layer is also given in bold next to the OSI name. There will, undoubtedly, be other uses of the lower layers not mentioned here, which will come to light as many Enterprises examine their requirements and operations.

The functions can be condensed to be performed by just one person or at maximum, up to seven layers of control as described in Sub-section 3.3. These levels were originally discussed as being relevant to Data Safety in an earlier edition of the Data Safety Guidance, (SCSC 2016). The principle here is that the OSI has been shown to be effective in simplifying connection between machines, therefore it should serve as a good model for communication between people, even if a machine is in between the communicating people, e.g. a telephone system, or radios. The functions of the OSI layers should, by such reasoning, serve as a model for layers of an Enterprise that wishes to communicate effectively within itself and with other Enterprises.

Note that each layer is provided in a single layer depth table rather than a seven-layer table, as those involved in an Enterprise with responsibilities for a layer of the LEDSM do not need to know much about other layers, so it makes it easier for them to identify and understand their role by using this format.

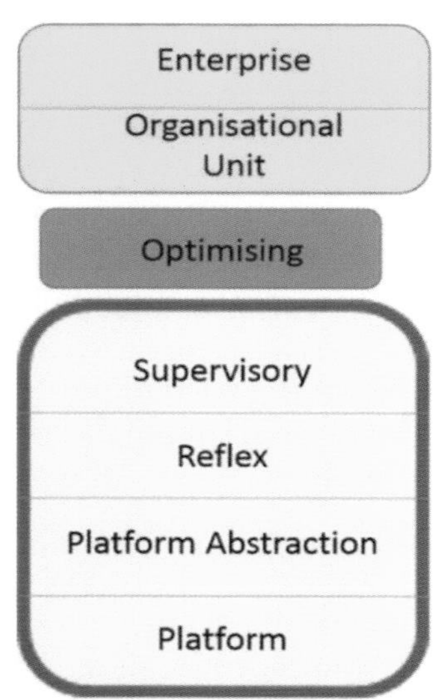


**Figure 8 ~ The Seven-Layer Data Safety Model**

Layer 7 – Application/**ENTERPRISE**

| | |
|---|---|
| The Application Layer at the top of the OSI model is what most software developers see; it is the layer closest to the end user. It receives information directly from users and presents incoming data to users. The application layer facilitates communication from the users' programs through the lower layers in order to establish connections with application programs at the other end. Web browsers are examples of programs that rely on Layer 7. | **This is the level of the Enterprise and the Enterprise owner. Depending on the application, this may be either an individual or committee. At this level, policy ownership is the main responsibility, and an individual should take responsibility even if the layer applies to a committee. Policy must be fed down in a comprehensible manner to all those who are the part of the Enterprise dealing with safety-critical communications. Policy must also be discussed and conveyed when finalised to the other Enterprise owners, the top-level personnel responsible for safety, so that each knows the data their staff can expect to receive, and what they are expected to deliver, though some Enterprises may sometimes be difficult to persuade to release data.** |

Layer 6 – Presentation/**ORGANISATIONAL UNIT**

| | |
|---|---|
| The Presentation Layer prepares and/or translates application data formats to network formats and vice versa. For example, it may provide encryption and decryption of data for secure communication. It is called the Presentation Layer because it "presents" data to the application or the network. | **In the LEDSM context, this is the layer where the policies of Layer 7, on how communication should be set up for particular applications, are turned into practices. Management intent becomes practical action. Basically, a translation to more technically safe processes. This, for best results, needs to be understood and agreed to by other Layers 6 in other Enterprises, through horizontal protocols. A good agreement will be a sound basis for preventing dire consequences in the future, should an emergency or disastrous system fault occur, whatever that system may be.** |

Layer 5 – Session/**OPTIMISING**

| | |
|---|---|
| When two networked devices need to "speak" with one another, a session has to be created; hence the "Session Layer". Functions at this layer include the setup, coordination (e.g. setting response timeouts), and termination of the communications channel between the applications at each end. | **This is the layer where action to instigate already planned procedures and processes occurs when an actual emergency arises. It will be the primary planning level too for those procedures and processes. In the ordinary course of events there would be a lot of feedback to Layer 6 and Layer 7 to ensure all issues were covered, but there would also be a lot of feed forward and expected feedback from Layer 4 to Layer 2 on how to practice these procedures. This must be done in advance of the emergency or critical situation so that the layers below react, in real time, appropriately. In the case of the Manchester arena bombing, that would be initiation of co-ordination of communication with all the emergency organisations and personnel involved.** |

Layer 4 – Transport/**SUPERVISORY**

| | |
|---|---|
| The Transport Layer deals with the coordination of the message transfer between application processes running on different hosts. It specifies how much data to send, at what rate, where it goes, etc. It splits the messages into smaller packets if necessary, and ensures that the messages can be reconstructed correctly at the other end. Common examples of protocols used at the Transport Layer are Transmission Control Protocol (TCP) and User Datagram Protocol (UDP). | **This is the part of operations that, when instigation of actions for an emergency has been received from Layer 5, will check that all communication links are in place. For instance, 'phone calls will be made to ensure someone is on the other end. If they are not, then alternatives must be found, so it would be a woeful response if all the necessary channels of communication and protocols to cover link failures, had not been put in place earlier by Layer 5 when the emergency network was in the development stage. This can be tried out in exercises, for example tests, such as: "What will the Layer 4 representative do if the primary mobile phone link between x and y is not functioning for whatever reason?"** |

Layer 3 – Network/**REFLEX**

| | |
|---|---|
| The network layer is responsible for forwarding packets from (and to) the Transport Layer; this includes sending through intermediate routers and network bridges that determine the best paths for the data transmission. | **This may be thought of as actual selection of communication devices as opposed to Layer 4 which is a more abstract view of devices. Layer 4 may state that a communication system with redundancy is required. Layer 3 will decide whether that is mobile telephones, land lines, radios, walkie-talkies, or computers. The protocols for in-service periods must also be considered. The question may be, "How many routes are needed to assure connection is maintained?". This is not trivial as proven by the 2002 Überlingen mid-air collision (2022), when three telephone communications were planned but because protocols were not thought through, especially where maintenance was concerned, all proved useless in the end to prevent the tragic crash. The protocols are the most important aspect of this layer, which would include prioritisation if multiple communication devices are intended to be available.** |

Layer 2 – Data Link/**PLATFORM ABSTRACTION**

| | |
|---|---|
| Most network switches operate at the Data Link Layer. It provides node-to-node data transfer between two directly connected network nodes; it also handles error correction for the physical layer. Two sublayers are included, the Media Access Control (MAC) sub-layer and the Logical Link Control (LLC) sub-layer. | **This is where the data being received is checked before it is acted on. As with electronics, the capacity of this should be adjusted depending on the level of risk and consequences involved just as error detection codes can be made more effective, but at the price of a measure of volume of data transmitted. In more-abstract safety-critical networks, such as urgent virus outbreak research, this layer may be seen more as the basic fundamentals of core interaction.** |

| | |
|---|---|
| | **In the example we look at later on, of a hospital in which a nurse had contracted a deadly virus, the hierarchy of layers would need to stretch down to the lower-skilled staff, who may, nonetheless, have important information. Lower-skilled staff would be classed as at this layer because it is at this level the data which appears important may not be and equally data which appears unimportant may be. The virus researcher would ask the lower-skill hospital staff who were good friends of the nurse who had died to clarify information they have. In the event it was those staff who were able to identify the most likely source of the virus. Note carefully therefore that just as an OSI based electronic system will not work without all seven layers, equally, LEDSM networks may be deficient without due consideration of all sources of data and information and in what form the information is delivered, even if on occasion all seven layers are vested in one person. Someone at this layer may be tasked with assuring management over and over again that all checks anyone could think of had been made. In the case of the Ariana Grande concert bombing this is the layer that checked there appeared to be no other terrorists and issued the statement that it was clear for emergency personnel, but sadly that message did not get through.** |

Layer 1 – Physical/**PLATFORM**

| | |
|---|---|
| At the bottom of the OSI stack is the Physical Layer, which is the electrical and physical representation of the system. This can include everything from cable type and configuration to optical wavelength and radio link frequency, as well as the layout and assignment of connector pins, voltages, and other physical requirements. When a networking problem occurs, the obvious thing is to check that all the cables are still properly connected, and the networking devices are powered. | **This layer will include regular checking that, for instance, standby mobile phone batteries are fully charged. It could also be particular people in the LEDSM context, as in the example given above for Layer 2, of a WHO researcher in Jakarta. That friend of the nurse who died proved invaluable in identifying the source of the virus. This may be where the intuition of communicating emergency and safety personnel proves so useful and why it is recommended that experience is captured for posterity, wherever possible, such as by databases being made available to all for whom they would be useful.** |

# 6 Developing Enterprise Links

## 6.1 Preamble

An Enterprise may combine or condense layers, especially at start up, and may add layers as it becomes a more sophisticated Enterprise. The initial state was illustrated in the diagram of the potential LEDSM type links that could have prevented the Chernobyl catastrophe, (see Figure 6). Victor Brukhanov's Enterprise, the construction of the Chernobyl Nuclear Power Plant itself, at the point in time that the snapshot of links is taken in the diagram, shows he is alone. Obviously, links are not fixed for ever in safety-related communication networks, just as electronic links between businesses using the OSI will not last forever, for instance when a business ceases trading and all its links to other businesses die, or when companies successfully bid for government projects when temporary links are established until the project is finished and accepted into service.

In this section we will look at how different links can develop for networks using LEDSM including Horizontal protocols.

## 6.2 Multiple Address Enterprises

In **Figure 3**, presented earlier, it was shown how an Enterprise wishing to link to an e-newsletter about viruses could fit that into a LEDSM diagram. The minimising of the connections in the diagram is useful, as potentially thousands of other Enterprises are connected to a network.

Figure 3 is a diagram of a fully mature Enterprise with seven layers, and connects to the ProMED journal via the Reflex layer. This is arbitrary for illustration only, and by convention within an Enterprise there could be essential safety-related News Services that connect to any of the layers, even the management layer at the top.

Of course, there will be many e-mails or news items on ProMED that do not relate to work that a particular Enterprise is involved in. All that means is that the messages of that nature will not be 'dealt with'. This is similar to the way messages on an electronic 'ring' network are received; if not containing the address for that network connection point, they will be ignored and allowed to pass on to whichever address they are intended to go. If appropriate, they will be allowed to pass up to higher levels of an Enterprise in legitimate receipt of the message, and any checking process for veracity and relevance done by people in a LEDSM network relates to this type of action. Also, normally one might try to filter the messages broadcast using keywords, or a similar technique, so that one does not have to read most of each message to check if it is worthwhile reading the whole thing.

## 6.3 Horizontal Protocols

It is not desirable to involve management in all the safety-related networking and data and information communication that goes on. In fact, it would be a recipe for disaster. This has a distinct engineering parallel, within the LEDSM, as it is normal in science, medicine and engineering, for instance, not to involve management with the engineering detail. So also in LEDSM, the management at the top level of an Enterprise is not involved in the day to day running of a company. Management will be looking at the strategic issues, issuing demands of staff, which equate to the vertical protocols in LEDSM. They will also be looking at who they need to network with at the management level, perhaps with regard to mutually acceptable protocols, which equate to horizontal protocols with other

Enterprises. Once agreed, it may require adjustment of the vertical protocols within their own Enterprise to ensure that agreed data and information flows exist between staff at the lower levels of the model who will need on many occasions to act instantly, and inform the equivalent layer in another Enterprise of their actions. Those involved in safety-related Enterprises at the higher levels should not be involved in the instantaneous reaction to safety-related incidents as they develop, but should have put in place the protocols that enable those reacting to know they are achieving the desired effects.

To illustrate how important this is, imagine if every computer and electronic system in a network used its own rules to replace the issues that the OSI seven-layer model deals with. Every device would be an almost amorphous mass as far as other computers were concerned, and it would have very little idea of how to communicate with other computers as they too would appear to be amorphous masses.

As the advent of the Universal Serial Bus, "USB", has proved, standards are important to simplify communication and connection. It used to be common to have to interface in different ways to each peripheral that was associated with a computer, the exact wiring being dependent on the manufacturer's whim. For instance, when connecting an Olivetti printer to an early personal computer it would require completely different wiring to the connections required for an OKI dot matrix printer.

Both the OSI and LEDSM work by standardising interfaces. What is important, where other Enterprises are concerned, is that the horizontal layers understand each other, because they have established protocols between them. This is not quite as the OSI electronic seven-layer model works, because in the OSI the protocols are fixed for almost all layers, (though there are sometimes small variations that overlap into other layers). In LEDSM, the Horizontal protocols are equally as important as in the OSI, but they are more flexible. This means a layer in an Enterprise may communicate to the same layer in another Enterprise with one protocol but to another Enterprise with perhaps a slightly less open protocol, withholding some information that is given out to others. This is like using different protocols to communicate with international newspapers, and other media, from those used with industry or government. The ideal is identical protocols, but this is not always practical; an advantage of LEDSM is that protocols may be public, so that the newspapers may be aware that they are not receiving all the information available but should understand what type of data it is that is being withheld. An example of this may be "Don't tell the press the virus is a Coronavirus, which it is, but do tell other disease control centres and research institutes that it is, because the press will run with some of them saying how dangerous those viruses are, whilst others are saying that we now have the immunological capacity to deal with these viruses so there is nothing to worry about".

Figure 9 illustrates both Vertical protocols that an Enterprise starts with and Horizontal protocols that develop with the Enterprises growing networks.

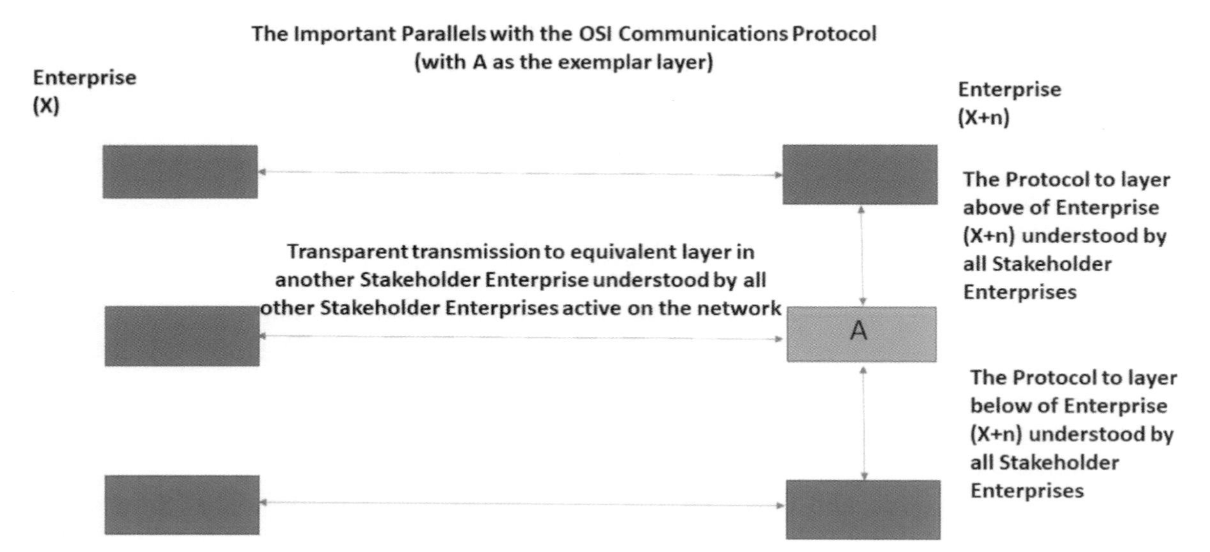


**Figure 9 ~ An Illustration of Horizontal Protocols**

## 6.4 Dependency Guarantee Relationships

The interface between Vertical layers in an Enterprise and between Horizontal layers to other Enterprises, must be governed by DGRs; Sub-section 3.8 described these in detail.

Although the term Guarantee is less commonly used, it is simply the complement of a Dependency; a dependency relationship is frequently not a reciprocated relationship, something is simply dependent on another thing in engineering, software development, etc. However, DGRs imply a deeper mutual relationship, which is what communication is all about, in as much that the dependent does not simply wait until it is supplied with whatever it is dependent on, but has negotiated a guarantee from the supplying layer that, whatever it may be, it will be supplied in a safe and timely manner. In other words, data or information is guaranteed both in accuracy and timeliness (ScienceDirect 2022).

In dealing with other Enterprises it should be remembered that the dependency and guarantee relationships within both Enterprises may need to be iterated until the relationship with the other Enterprise defines the data to be supplied and received that will maximise risk reduction for both parties. That will then formulate the finalised protocols both vertically and horizontally. This is not a trivial point as the communication protocols can easily be tailored to individual Enterprises, something that is not a general feature of the OSI.

## 6.5 Data Passed Within An Enterprise

Unlike in the electronic application of OSI, it is not the purpose for all information to go up and down the entire seven layers (minus the checking data) as in when one sends a picture to a friend over the internet. For instance, some direct horizontal communication with other Enterprises may occur between higher levels of the model, e.g. management to management. However, in order to maintain the safety integrity of the Enterprise and those with which it communicates, an appropriate level of detail of what the communication comprises would be sent vertically down or up the layers, as appropriate and in accordance with the agreed protocols. A simple example from the world of electronic communication is that, in sending a message, one does not wish the cyclic redundancy check that adds check data to be a task of the sending user to code, and one does not want the receiving user to have to decode each message to be sure it has not been altered in transmission. That stripping of check data electronically is the equivalent in LEDSM of allowing management to control what data is released at what level, if and only if the protocols management wishes to be in place have been carefully thought through and agreed as appropriate.

# 7 Use of the LEDSM in Emergency Service Response Planning

## 7.1 Preamble

Poor Emergency Services communication, after the Suicide Bomb Explosion at the Ariana Grande Concert in Manchester, is considered to have led to avoidable deaths. The following is an example of how LEDSM would have been useful, had the networks involved been developed in advance using LEDSM's logical processes. Just a small number of well thought through connections would have changed the face of the outcome, resulting in fewer deaths and greater confidence in the competence of the emergency services. In the event, 22 people died and 112 needed hospital treatment for their injuries. Many waited a long time in pain for help. The failures of the emergency services in responding to this suicide bombing are quite shocking and deeply disturbed the relatives of those who died or were injured.

The next section will look at more-complex problems, but this section is intended to build on understanding, so that a full grasp of the issues can be established before venturing out onto more problematic issues. Working through examples increases one's ability to meet any challenges in the future, much as Engineering students may work to solve scores of electrical technology questions, integral equations, Heaviside Step function problems, Laplace Transforms, etc. By doing so, they ensure they understand the subject sufficiently to take a professional approach when they start their careers, and can solve final exam questions too, of course...

## 7.2 Critical Issues

The effectiveness of emergency responses could be increased by planning using a LEDSM tool, *a priori*, to analyse possible situations and responses and what needs to be changed to reduce the risks identified. Here we look first retrospectively at what did go wrong. As with many safety-related incidents, the best way to analyse what went wrong with communications in this case is to look at the timeline of known emergency actions. Table 1 is derived from an article, "The Lost Two Hours" by David Collins, in The Sunday Times edition of 25th July 2021 (Collins 2021).

**Table 1 ~ The Ariana Grande Bombing Emergency Response Timeline**

| TIME | EVENT |
|---|---|
| 10.31 | Salman Abedi detonates suicide bomb in the "City Room" foyer. Public dial 999 |
| 10.41 | Armed police officers arrive = Armed Response Unit (ARU) |
| 10.44 | ARU realises there are no other terrorists. One issues the request, "*We need paamedics like f**king yesterday*" |

| TIME | EVENT |
|---|---|
| 10.45 | Andy Berry, Duty Liaison Officer (DLO) for the Fire and Rescue Services, without evidence, takes action to "protect" firemen by ordering them *not* to approach the building. |
| 10.47 | Inspector Dale Sexton of the police declares "Operation Plato", effectively forbidding rescue services and paramedics from entering the building, again with no evidence that it was an attack of many terrorists, as in the Mumbai hotel attack. |
| 10.50 | A paramedic assigned to assess situations before others enter, arrives in the City Room, but does not treat anyone. |
| 11.12 | Inspector Sexton declares to his control room that he is "*reasonably satisfied there are no other terrorists*", but he fails to call off Operation Plato. |
| 11.17 | Seriously injured John Atkinson is carried from the City Room to a casualty clearing location. There appears to have been little effective triage action, which should have identified John Atkinson as highest priority. |
| 11.48 | John Atkinson finally cleared to be taken to hospital but dies from loss of blood. |
| 12.37 next day | The fire engines that could have provided stretchers (makeshift ones had been used by emergency workers and volunteer members of the public to evacuate casualties), finally arrive at the concert arena. |

The critical issues then are:

1. Inspector Dale Sexton took actions that implied there was evidence of other terrorists when declaring Operation Plato. There was no such evidence.
2. DLO Andy Berry directed fire engines to a place three miles from the scene of the bombing to protect them from the risk of there being other bombs or terrorists. There was no evidence that this was necessary.
3. Inspector Sexton's phone was constantly engaged, so when DLO Berry tried to contact him for better information he got no reply.
4. The ARU's knowledge that there was clearly no evidence of other terrorists, did not get through to either DLO Berry or Inspector Sexton.

### 7.3 How Could Using LEDSM Have Helped?

Firstly, there would have been an analysis during development of a LEDSM that showed the communication path between DLO Berry and Inspector Sexton was a critical link. To do this one could use WBTR[24]. In Section 10, we will look at WBTR using a worked example. What a thorough WBTR analysis would do is expose where, for instance, duplicate systems are needed. It would therefore have become not only a requirement for a dedicated hotline between key people making the decisions, but also for at least one redundant link between those persons to ensure communication can take place. This, on the LEDSM Enterprise diagram illustrated in Figure 10, could be inserted at Layer 2, which in electronic terms is the Data Link checking level.

Analysis at this stage thereby, would have meant that the protocol vertically down the Fire Services Enterprise from DLO Berry and, similarly, vertically down the respective Enterprise from Inspector Sexton to the lower 'automatic' levels, would demand that at least two and, probably more-appropriately, three communication channels were always

[24] What/Why Because Therefore Reasoning

available between their two Enterprises. That way if there is a fault, a workaround is in place. The two officers might both be thought to be at Layer 5, the Supervising level.

As hopefully is clear, the problems that may occur need to be understood by the Layer 5 persons. Protocols need to be in place to ensure Layer 2 gets the message across to whoever needs it. In addition, the protocols allow those at lower levels to check the message they receive from the operations level of another Enterprise, (i.e. like the checks performed in electronic systems on incoming messages) by asking the question, "Can you confirm that other terrorists are known to be present?". In fact, the Police Enterprise commander might then have understood his mistake and declared that other terrorists were not present, and hence the firefighters with stretchers and others could have entered earlier.

To summarise, at the concert, a major problem identified by the inquiry was the lack of communication from the hall, where the murders occurred, to other emergency service personnel. This was due to two problems hindering the communication:

- The first was a lack of protocols between the front-line worker in the hall and the leaders of the organisations. The protocols need to be in place so that those who call the tune are aware of the facts before they call that tune.
- The other was a lack of communication systems between the front-line worker and other levels within that Enterprise, any of which could have communicated to the other emergency services.

So, the analysis using LEDSM, for the front-line worker entering such a building, shows that there must be an ability to communicate to equivalents in other emergency services as well as those senior within the organisation. It may be that such designated workers could be linked and have their own protocol for who goes in first, what is communicated to the other two or three personnel and when they enter the building too. In Figure 10, the term 'Flash Message' is borrowed from the military for messages that must be dealt with instantly, not within minutes or hours.

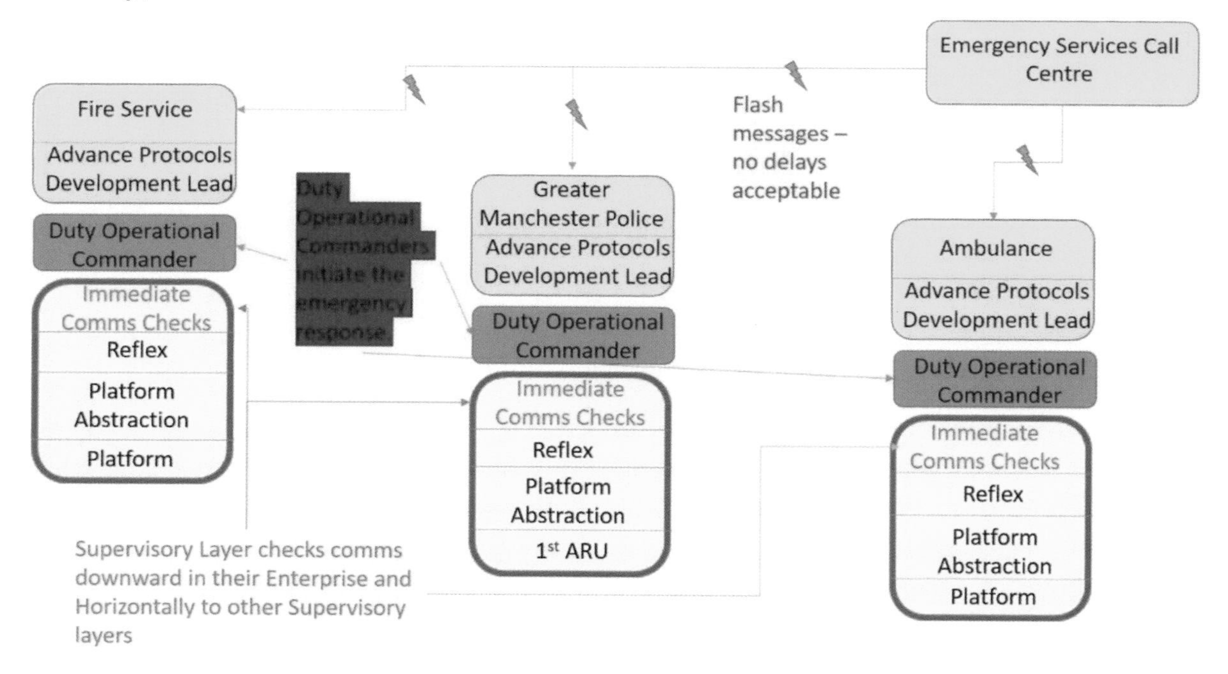


**Figure 10 ~ Using LEDSM Could Help Identify Potential Communication Failings**

This is a similar problem to what went wrong in the sky over Überlingen when two aircraft collided, partly because a telephone that should have been used was being upgraded, and so a warning from one Air Traffic Control centre to another could not be made. On that occasion the available communications would have been adequate, but nobody had thought through what could go wrong during maintenance. Three telephones should have been enough. Some extra communication channels, simple but effective, should have been in place for the Manchester incident.

As many people can die in a terrorist incident as an airplane crash. As an example, one may compare the, at least, 174 deaths during the Taj Hotel attack in Mumbai with the 157 passengers and crew who died when the Ethiopian Airlines Boeing 737 MAX crashed. There is no excuse for unpreparedness in operating with single communication paths, wherever threats to life can exist, when multiple communication paths are so plainly required to reduce risk.

### 7.4 Summarising Key Points

As a guide to the use of LEDSM, DGRs and, where necessary or useful, WBTR this paper cannot go into too great a detail, but shown here are two requirements that would have, had they been included, brought relief to those suffering after the explosion and probably also saved lives. Both emerge from the use of LEDSM. A brief summary of what has been shown is then included:

**The Advantage of Multiple Links:** There should have been a hot line with a redundant extra line in case of failure of the first method of communication, between the ARU and Inspector Sexton and DLO Berry. It is vital that data is provided in a timely and trusted way. LEDSM model development would show the communication links that would be necessary and enable establishment of protocols between Enterprises (three in this case, the Fire and Rescue services, the Police command centre and the ARU). Those protocols would establish who expects what information from whom, and within what time parameters, both within each of the Enterprises and between the Enterprises. Each Enterprise would then have its own protocols to describe how those within the Enterprise were to subsequently operate once inter-Enterprise communication had established agreed actions. It was particularly sad that the ARU confidence that there were no other terrorists was not conveyed quickly.

**The Lack of Wisdom in not Defining Protocols:** The “Operation Plato” protocol was instigated in the light of events in Mumbai when the Taj hotel was attacked. It had not however, been communicated that three minutes earlier the ARU had decided there were no other terrorists present in the building. Evidence has to be the basis of such decision making. One advantage of LEDSM is that when new events do come along that alter thinking, scenarios can be re-imagined so that should the event occur a second time, those involved are prepared. The LEDSM model, when implemented may be modified after an event when reactions were considered inadequate simply by discussion of how to change protocols and add links to other Enterprises in the model. Decisions are formally documented by the protocols preventing, thus, any human errors later in recording what should be done or forgetting to pass on the decision to later incumbents of the post.

**This Example Illustrates:** That LEDSM provides an easy visual way to develop communications, even when those involved are not safety-critical experts themselves, (but see Sub-section 7.5). The principle at the inter-Enterprise communication level, (Layer 2 in this case) is incorrect communication risk reduction.

The passed down protocols are an essential part of the responsibilities of those charged with providing safe environments for their staff and the general public. Through the subsequent reasoning around the protocols, the necessary minimum multiple channel network required to assure timely and correct communication will be confirmed and developed. To not apply a formal method that has, as an output, an easily understood diagram, as LEDSM does, is extremely risky. We need to refine our use of technology, much as we improve the technology itself, by striving for the best and safest products and processes.

### 7.5 Qualifications to Support Emergency Response Preparation & Implementation

It is well understood that the response of the New York Fire Department (NYFD) to the attack on the twin towers of the World Trade Center was conducted in ways that exemplified courage and leadership. One of the prominent characters in the tragedy was Battalion Chief Orio Palmer, who led the team of firefighters that reached the 78$^{th}$ floor of the South Tower just before it collapsed. There were radio communication problems for the emergency services at this incident, and recordings of Palmer have helped understand the problems. Palmer himself had long recognised the problems of communication, especially in tall buildings. He had an associate degree in electrical technology and had written many articles about the problems. He has been described as being one of the most knowledgeable people in the NYFD on radio communication in high-rise fires. He had even authored a training article for the department on how to use repeaters to boost radio reception during such emergencies. This shows the importance of having SQEP in a position to apply themselves to problems concerning the effectiveness of the communication plans and system design when dealing with safety and emergency issues, especially communication, as mentioned in Sub-section 3.10 with regard to Chernobyl. Although trivial in terms of complexity, it is nonetheless true that the hypothetical application of LEDSM, to the Manchester emergency response preparations for such an incident as occurred at the Ariana Grande concert, would have resulted in a much better outcome and especially if SQEP were involved.

## 8 Multiple, Flexible and Dynamic Network Protocols and DGRs

### 8.1 Preamble

As described already, an Enterprise can link to any number of other Enterprises and in doing so should engage in the establishment of horizontal protocols. These protocols describe how the linkages to other Enterprises will work at the various levels. Clearly, if there is only one person in the Enterprise then all links are to him or her but obviously that link may be to a multi-layer Enterprise. That then will involve some negotiation as to whom in the bigger Enterprise the link is to be made, and it may be different levels are linked to when network connections to multiple Enterprise are made. It may also be the case that a single individual will link to multiple layers within an Enterprise, where the communication is to establish cause and effect or other enquiries. We next look at this aspect.

### 8.2 An Example from the Health Sector

Figure 11 shows the possible connections a WHO front-line researcher may have had in the hunt for a particular outbreak of a virus that, it later transpired, had come from infected chickens. The figure is based on the health research work of Dr Gina Samaan, see Figure 12, of the WHO Jakarta office at the time, since then with the US Centre for Disease Control' influenza division.

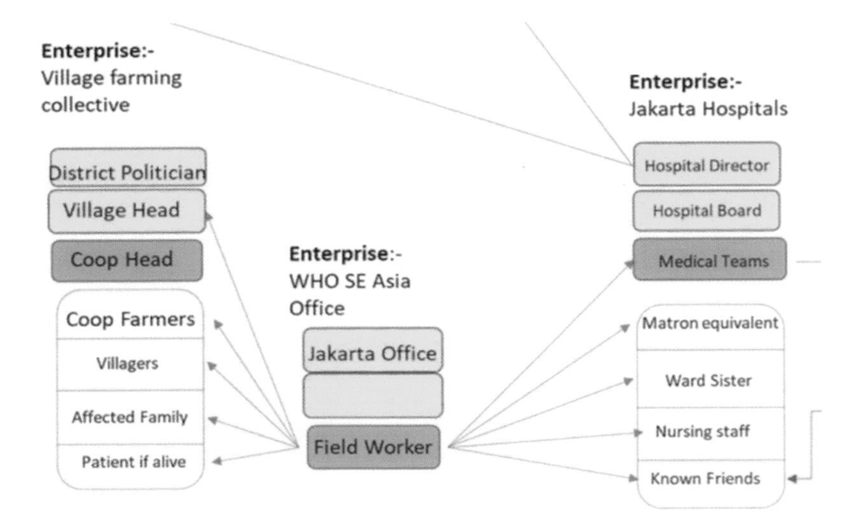


**Figure 11 ~ The Potential Links of a Field Worker Researching a Viral Outbreak**

This incident was previously discussed in a lecture to the 2021 Safety-Critical Systems Symposium (Hales 2021). It eventually transpired that a nurse, the first victim, had not picked the virus up from the hospital where she worked, nor from her village, both of which were high up in the list of likely places, but from a Jakarta wet market where she shopped for chicken. The virus had likely jumped species to infect her probably because in wet markets, viruses can easily move between the live species through breathing the air, contamination from defecation, or direct contact. A virus then may combine with viruses already present to create a more deadly virus. Chances are that at some point, that virus will infect a human and if it is a serious infection, it only requires one more step, the ability to transmit freely between humans, to become a pandemic.

**Figure 12 ~ Dr Gina Samaan (Doctorate in Epidemiology)**

Not all possible links of each Enterprise are shown in Figure 11, but the lines from the hospital director can be expected to link to many health-related Enterprises and the Jakarta office of the WHO would of course have many other links, but this can be thought of as WHO researcher Gina's LEDSM diagram for the instance of this particular virus outbreak. If it is drawn out like this, should Gina have tragically died from the infection due to the close contacts she bravely made, then at least the diagram would show a replacement worker where research was up to in an easily understood way and contacts could be re-established.

The contacts of the researcher and the protocols for the hospital Enterprise are not trivial relationships that can easily be imagined, but LEDSM diagrams assist understanding.

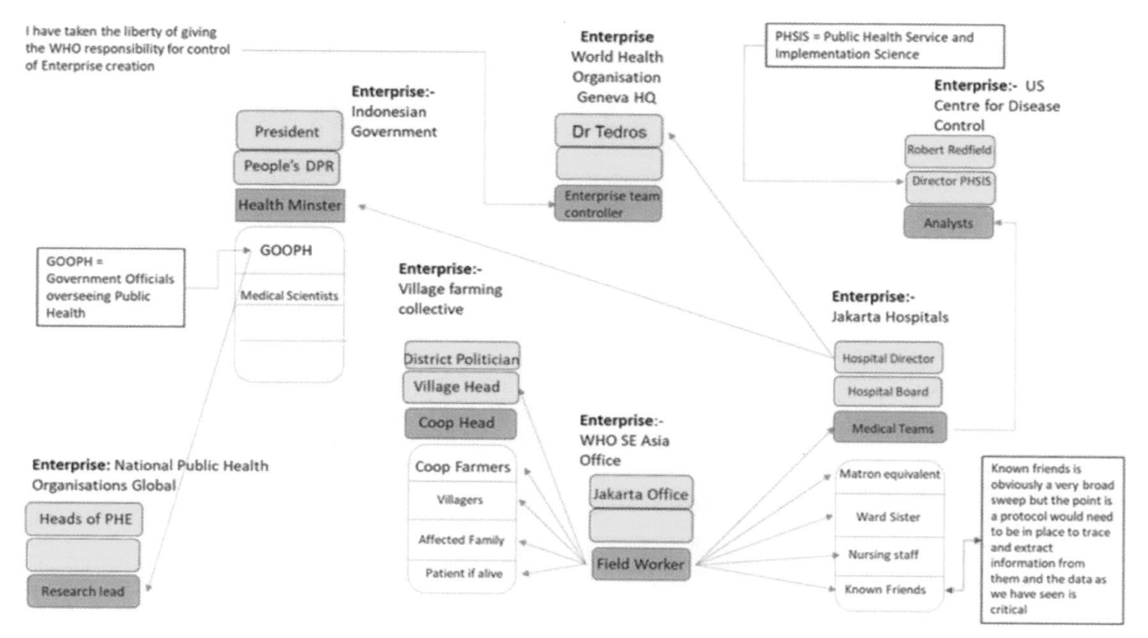


**Figure 13 ~ Snapshot of a WHO Virus Outbreak Researcher's LEDSM Connections**

Figure 13 shows the increasing level of complexity as we expand outwards to see connections other than Gina's direct connections. There is the potential for conflicting interests. If planned properly in advance, they can be resolved simply as engaged stakeholders see what ought to be their obligations. This is again where DGRs become useful in that they help define protocols. This method of defining in advance via use of LEDSM diagrams would enable, in this case, quicker alerting of the world to a virus outbreak. Interestingly, in this case, Gina has to talk to people at all levels because it would be beyond the powers of say, the sister in the ward, to have knowledge of all the questions that Gina may ask her staff and then give a response to Gina on their behalf.

Hence the protocols in place for the hospital would involve the sister, at a Health and Safety briefing, having alerted her nurses that, at some point unknown at the time of the briefing, they may be required to go into detail about their personal lives and habits and what they know of the personal lives and habits of their friends and other contacts, should a virus outbreak occur. This equates to a basic track and trace for when a sophisticated "Mobile App" enabled tracing may not be possible. The situation would be similar in the case of the village Gina visited too.

## 8.3 Getting the Balance Between Safety and Security

The classic example of conflict of interests is in determining the right balance between safety and security in a nuclear plant. It is literally vital that in the event of an accident, external organisations, such as the emergency services are notified as quickly as possible, which means automatically. However, it is also equally vital that the nuclear plant is protected from terrorist or hostile government hackers. In this case of pursuing a virus outbreak, the hospital will want staff to cooperate with the WHO researcher, but in general will not want staff to talk about their work for confidentiality reasons. Planning in advance using LEDSM will ensure the researcher has access to the staff on the ward, without a long debate at the management level of the hospital about whether or not all staff or just some senior staff should be allowed to talk to her, and whether lawyers need to be present, followed by a subsequent negotiation with the researcher. A sense of urgency is of the essence in a viral outbreak as in any other emergency, by definition. Using

LEDSM, most if not all internal and external protocols will have been agreed well in advance of the virus outbreak and briefed to staff.

Looking again at Figure 13, it is a snap-shot of what a LEDSM diagram for the WHO researcher would have looked like in the expanded form that shows many more stakeholders dependent on the data Gina uncovers being transmitted. The Enterprise for the village is a template exemplar as it is unnecessary to get every village head in Indonesia to sign up to protocols, just the one dealing with the issue at the time. As stated before, these relationships can be dynamic in slow time, being re-examined to ensure they are at maximum efficiency as and when changes arise within or outside the Enterprise.

There also may be a need to have separate diagrams if parallel but different issues need to be dealt with, so it may not be all that the researcher looks at. It may be replaced by a similar diagram of 'planned in advance' protocols with other Enterprises, for instance another hospital in another country where different protocols were in place if another outbreak arose while a first is unfinished. Figure 11 also shows how important the very base of the Jakarta Hospital Enterprise is, as only one of the victim's fellow nurses held information that was the clue to where the outbreak had arisen from, telling the researcher where the victim would, usually and frequently, go to pick up meat before returning to her home village.

If everyone involved in emergency and safety work had close to hand a diagram of their evolving links to other Enterprises, it could reduce quite a number of errors in accident, disaster and emergency planning and handling.

When a large number of connections are going to be involved, e.g. in a global pandemic, different protocols may need to be in place for different stakeholders.

# 9 Meeting and Adopting the Eleven Principles of Safe Software Design

In this section, we will look at eleven technical principles of best software design, as espoused by Martyn Thomas in "A View from the Stern" (Thomas 2005), but here adapted to the LEDSM method of safer data network design. That safety-related data and information issues slip so easily into the same principles as safety-related software issues, illustrates how important the task of improving data and information safety transmission is. Safety-related software is constructed in a much more formal way than other software to ensure failures requiring re-booting, or bug fixes in slow time, which are tolerable for software with less associated risk, are not included in the safety-related software program. Similarly, data and information safety systems must be constructed with a greater degree of formality than currently, in order to ensure they do not fail when they are called on during an emergency.

In many cases, simply replacing the term 'software' by '**human-machine combined data and information networks**', or words to that effect, provides the initial principle, though more is discussed. It also applies to human-to-human data and information networks as we have already met in the shape of the investigations of Gina Samaan into a virus outbreak at the Jakarta hospital.

1. The difficulties of designing a safe 'human-machine combined data and information network', flows from the increasing complexity of the problem space, as members of the network are added, contrasted with the need for simplicity to retain comprehensibility. A pertinent observation often ascribed to Albert Einstein is, "*Every system should be as simple as possible and no simpler*". Modern engineering has given us an array of devices to utilise in

communication systems which are, by and large, user friendly, for example the mobile phone, the laptop, the megaphone (still important for tsunami warnings). These are the equivalent of hardware in safety-related systems that require software. The data and information we transmit and receive over these systems is the equivalent of the software. As we get to know more about our world, and people have more space and time to enjoy it, negligence that leads to suffering becomes increasingly taboo. Hence the complexity of the space naturally expands to accommodate demands for almost 'perfect safety'. LEDSM helps to keep networks comprehensible to all involved, regardless of whether or not they are safety professionals.

2. 'Human-machine combined data and information network' transmission errors are systematic. In this case the similarity is nuanced because if LEDSM is used, the errors do become systematic, programmed in by the protocols; whereas, normally, one would consider human communication errors as random. Like software, data and information emergency network faults may only be detected years after the inter-personal connections, and computer connections, are commissioned into service, sometimes only when the actual emergency the system of data transmission was designed for arises. This is similar to how things are for software in complex safety-related systems, where many more paths, of the order of millions, may exist with most never being exercised until many years of usage have passed. Only when an emergency occurs does the fault become apparent. The difference is that data and information safety-related systems involving humans are generally less complex to analyse but when called on are required to perform a similar task, saving potentially many lives. LEDSM is an analysis tool one could use. In emergency and disaster prevention communication, some of the problems are equivalent, in terms of being problematic to test, as is testing every path in a complex software program. Risk is many faceted and nuanced, especially by context. To commercial Enterprises, it may be primarily a function of environmental impacts and economic costs. Where humans in the Enterprise are involved as part of the system, as in the communication networks referenced here, resilience under pressure and the ability to adapt must be included. Where terrorism or military conflict is involved, networks must take account of threat, vulnerability, and consequence. The probability of failure is more appropriate in the context of reliability of machines. The risk that the environment can change continuously, especially in on-going emergency situations is one which must be addressed but with a little thought in advance, a degree of preparedness can be realised. One way to reduce the risk is by making the network more robust to failure.

**Ambiguity in language.** Solved by a protocol governing format of communications as in Hales (2021).

**Paucity of data.** Solved by establishing, during emergency communication development, what data will be expected and what will be delivered.

**Delays in transmission due to bureaucracy.** Solved by management instigating protocols that permit the personnel or systems of the lower four layers of the LEDSM model, whether compressed or not, to act in an emergency autonomously, having been trained or **developed well, in advance.**

**Poor design of the network**. Improved by allowing all members of the network to see visually their connections with LEDSM diagrams, and thus opening up the action spaces to intellectual assessment of other possible connections.

**Over-confidence encouraging resistance to any declared need for communication**. Solved by analogy with electronics, such as extra redundant paths for communication and Cyclic Redundancy Check equivalents.

**An absence of necessary nodes in the network**, such as having a network dealing with virus experts except the one person that knows more than anyone else, the expert on the particular type of virus. In the case of the Ariana Grande concert bombing and the Überlingen air collision, a lack of sufficient effective communication channels was significant. The addition to a multi-channel radio system of mobile phones, and dedicated mobile phones, reduces the risk of important messages not getting through. Figure 14 illustrates the opportunities the Police Commander could have had, if communication had been thought through a little more, to send a message that there were no other terrorists present in the building. An advantage, for instance, of a dedicated mobile phone is that if pressure of the situation means one person cannot receive radio messages, as actively communicating to others, messages can be left, which is not a feature of many radio communications. All these issues are more apparent by diligent LEDSM planning. Note, it is not suggested this is actually the solution that should have been adopted, it is simply here to demonstrate the process.

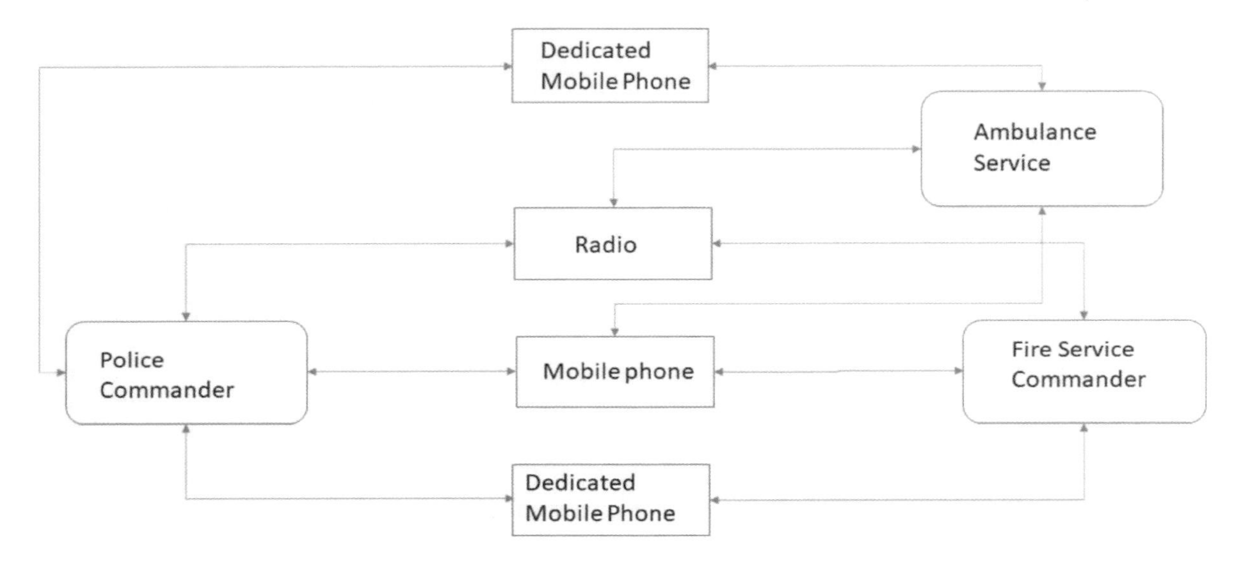


**Figure 14 ~ How Increasing Emergency Communication Channels May Reduce Risks**

3. Abstraction is the only way to conquer complexity. This is what the LEDSM model is designed to do, initially by providing a tool to envisage data and information paths, but also providing abstraction techniques, such as characterising an e-mail news source, provided to specialists that have signed up to it as a subscription service, as a single Enterprise rather than trying to draw connections to all relevant members. The abstraction is in trusting the machines to some extent, thus avoiding analysing the complexity inside them and relying on redundant systems to solve any concerns over reliability. Edsger Dijkstra stated, "*...the purpose of abstracting is not to be vague, but to create a new semantic level in which one can be absolutely precise*" (Dijkstra 1972). LEDSM enables such precision through providing diagrams with traceable connections rather than lists of names which may mean nothing to a third party invited to peer review safety-related data and information connections, and which are unlikely to be checked individually, or be difficult for a new person in the role to

understand. LEDSM provides an effective training tool for those new to a role by visual explanation of communications necessary.

4. In data and information transmission, low defect rates and low cost are achievable together. Just as in safety-related software, though, the cost of an accident can be enormous in terms of lives lost and injuries — and of course compensation for the victims, such as the millions of dollars paid out by Skyguide in compensation for the Überlingen air collision, or the billions paid by Boeing for the two Boeing 737MAX aircraft crashes (Schaper 2021). However, just like software development, early intervention can be crucial in reducing cost. The application of a visual modelling process, such as LEDSM, well in advance of emergencies assists this early intervention requirement. Also, there is the advantage that remodelling is incredibly cheap but simple to observe the effectiveness of, as in many cases one may simply add an Enterprise to the diagram and then make the necessary connections to provide even greater safety. Removal of connections to an Enterprise is equally simple.
5. Maximise Cohesion and Minimise Coupling is a maxim for modular software development. This applies with the LEDSM too, since obviously nobody involved in safe data and information transmission should want to be receiving information they do not need, nor should they want to be providing irrelevant information to others. Also, those legitimately involved should not want to share with others outside the protocol that has been agreed between those within their Enterprise and those in other Enterprises with whom protocols have been set up. The LEDSM technique enables easy discussion among members of an enterprise as to what they feel they are lacking in information and from what source they may be receiving too much information, hence enabling connections to be made or broken with protocols kept in place or modified by agreement.
6. Maintaining Traceability: LEDSM enables the tracing of connections to requirements, since formal protocols within an enterprise determine what type of connections should be made and why. Those connections may be discussed as to how they fulfil requirements of the top level of the Enterprise at any of the regular reviews.
7. Maintain Version Control: This is a hugely important technique in software development and maintenance. It is also true in the use of LEDSM. It is important that when a connection, with its agreed protocols, is established, changed, or removed, the other parties are informed of that so that they have an updated version of what their network looks like. An example of this would be a Disease Control Centre dropping a link to another Disease Control Centre at the behest of their government. This could be disastrous for some if the second Disease Control Centre thought they would be informed of a disease outbreak, but instead only learned a week later of rumours from any media outlets picking up on an internet rumour.
8. Testing is an experiment: Testing the data and information network is not a bad idea, "*Experiments are most effective when they are designed to disprove a hypothesis*" (Thomas 2005). Basically, this means that any network an Enterprise decides to build should, when tested, utilise tests that try to demonstrate that the connections they have built will NOT work. It has been shown time and time again that testing to prove a software system works simply hides the faults because the thinking necessary to find faults is not done when all one wishes to prove is that what one has constructed works. This is where investment helps. Modern disasters, such as the two Boeing 737 MAX accidents, were down to a lack of rigorous testing, dismissiveness towards safety concerns, and the process of allowing self-certification of the aircraft by the

manufacturer; the equivalent, in this case, of designing the tests to prove it works, instead of trying to show where there are faults. Because emergency systems are so rarely used, reliance on statistical data, e.g. "*the one exercise we did worked well*", is not a very sensible path to follow. Regular rehearsal is necessary, as is network review. Exercises should include trying to prove it will not work, such as setting up radio links to brick lined stairwells in skyscrapers, a problem briefly described in Sub-section 7.5.

9. Safety arguments depend on sound languages. LEDSM is a form of visual language. It reduces the likelihood of: tardiness in communication, a lack of preparation for emergencies, misunderstandings within Enterprises as to intent, an absence of a sense of urgency, and misunderstandings across Enterprises as to what is and what is not available data and information. In that sense LEDSM performs the function, (to a certain extent in its own field of network communication), of a safety-critical software language such as SPARK Ada. It reduces the amount of data that can cause problems and increases the clarity of data that is passed.
10. Probabilistic Risk Assessment (PRA) has limited value in this context: Such an analysis relies on a belief that all causes or modes of failure have been taken into account. As stated in an earlier section, we can use the adaptation of WBA, WBTR, and this is illustrated with examples in the next section. The realistic way to reduce risks is to go through the process of assuming there are causes or modes of failure that will not have been taken into account. We can then use simple basic safety principles to analyse the needs of the system in question. Many software system safety cases in the past have used PRA and it often proves inadequate for the job, but it is especially not appropriate for data and information safety, as the variations on data communication may include language nuances between nations or a lack of understanding of the subject being communicated. An example of that can be found in some management circles, where they expect engineers to just get on with their work. This is no shame on the part of management because we cannot all be expected to know everything that occurs in an Enterprise. The risk though is that management has not fully understood the disastrous consequences of the communication system failing, so it is important to ensure that LEDSM protocols employed between layers in an Enterprise convey the urgency and importance of the work to management that the lower layers of communication will do in an emergency situation. There is also often a lack of safety back-up systems, for which probabilities might be thought of as miniscule but nonetheless, because we are dealing with human beings in systems can very easily occur. Measuring human reliability is no easy task. Much better then, to establish protocols which permit autonomous action by the lower levels of the LEDSM model than expect instructions from management at the time of the emergency.
11. Standards should not set unscientific targets: This means in the data and information safety context that some things can be guaranteed, but the idea that assurances can be provided that nothing can possibly go wrong is a red herring. The aim of LEDSM is to reduce risk, not imply that by adoption all possible problems are eliminated. For instance, the currency of the model an Enterprise holds should be reviewed every six months, say, to ensure scientific advances are taken on board and new Enterprises reviewed to check if there is potential benefit in communication with them. That does not mean that somewhere there will not be an Enterprise that is not on the network the Enterprise is using, that would have been useful, but LEDSM does reduce the risk. In addition, when reviewing networks, it will be important to keep in mind what the purpose of the

network is, e.g. to keep all stakeholder parties aware of the emergence of a new threat. The key is to act quickly when an obvious omission has occurred and to set protocols that are scientific, i.e. rational and practical, but exclude political decisions that reduce safety. This includes not hiding information from those who need it. The same applies to other emergency and safety situations discussed, such as emergency service responses to a terrorist bombing. In planning the network, start by 'observing' the history of such incidents and the problem space in order to deduce the initial rational protocols.

# 10 Using WBTR to Establish Safer Designs and then Develop a LEDSM

## 10.1 Introduction

WBTR is a technique derived from WBA but used as a development tool for safety-critical communication, rather than as an accident investigation tool. It is considered to be useful to reduce risks in data and information systems by assisting in the identification of what communication systems should be in place, and with who, in order to ensure safe responses to emergency situations and others where lives are at risk. It needs to be used with LEDSM in order to maximise effectiveness.

The basic idea is to look at the design of the system you intend to construct, e.g. in a typical HAZOPS it would be a chemical plant, and ask key words about the processes. In HAZOPS a question might be, "What would happen if more of substance x flowed into a vat than intended". In WBTR safe communication developments, the brainstorming should look at any proposed communication systems and ask questions such as, "What would happen if the 'phone link was out of action?" and, "Why would Firefighter A not be able to hear the Fire Marshall's command to abandon the building?". That in turn encourages participants to state formally why, for example, "Because of radio interference, the command may not be heard clearly enough". Then the brainstorming moves to 'Therefore' reasoning, e.g. "Therefore, there should be two or three Walkie Talkie channels, signal boosters, and a mobile 'phone given over to the Fire Marshall exclusively, so that interference can be dodged". The analysis may conclude there should be a spare walkie talkie held by another member of the team to ensure that a breakage, or the death in an inaccessible place of the walkie talkie holder, do not prevent the team hearing the Fire Marshall's communications. In addition, "All firefighters should carry text 'phones or pagers through which the command may pass via civilian channels in case all radio channels are blocked", may also emerge as a solution. Some basic rules, such as the use of multiple machines to back-up chosen communication routes are necessary. Figure 15 is a simple illustration of the use of three channels between two persons to give this basic, but all too often overlooked, risk reduction.

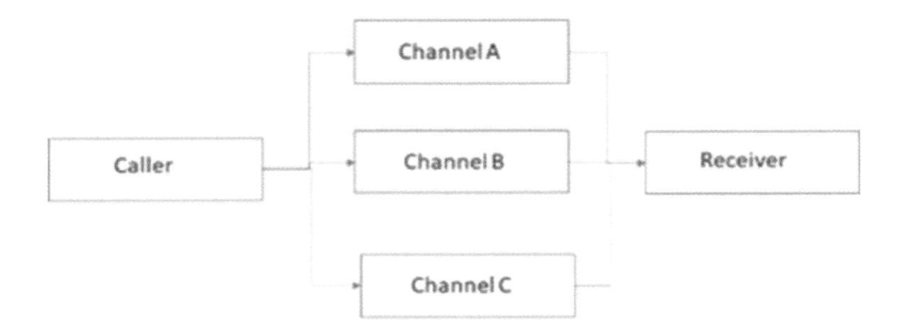


**Figure 15 ~ Multiple Channels Illustrated**

The first thing to do in building a safe system with potential communication problems is to identify what data is safety related. It may not be obvious, hence the use of WBTR will help. One may think the important thing is that each person knows their job but, frankly, that is just about what happened at Chernobyl, a dire communication failure that was briefly examined in the context of this paper, in Section 3.

## 10.2 An Example of the use of WBTR in the Überlingen Mid-air Collision Context

### *10.2.1 Preamble*

The following illustrates how the collision of two aircraft could have been prevented by using WBTR and LEDSM to assure the effectiveness of Zurich Air Traffic Control's night-time operations. The two techniques are effectively iterative since each time one reasons that an additional provision must be made, this naturally invites one to revise the associated LEDSM diagram connections, and that in turn will help in reasoning through what could possibly go wrong. The ease with which LEDSM diagrams can be changed makes this much simpler than one may at first think. Where safety-related communication networks are involved, the use of WBTR, along with other techniques, may produce improved Hazard Analyses.

Considering the lack of reliable probability data for human failures, WBTR may have advantages, since probability is not considered in WBTR. Only the brain-storming session comes close to assigning probabilities, but then only when thinking possibilities through. The advantage to bear in mind with WBTR is that humans are involved in most of the communication systems for which LEDSM will be used as the communication development tool. Humans are fallible. Human involvement means a high degree of complexity is inevitable, since we are complex beasts. We are prone to forgetfulness, oversight, panic, and excessive confidence in things we know less about than we ought to.

Figure 16 is a diagram of the systems involved in the air collisions at Überlingen. The first actions in developing a reasoned argument using WBTR is to draw a system diagram with connections, as illustrated. Once that is done the brainstorming can begin.

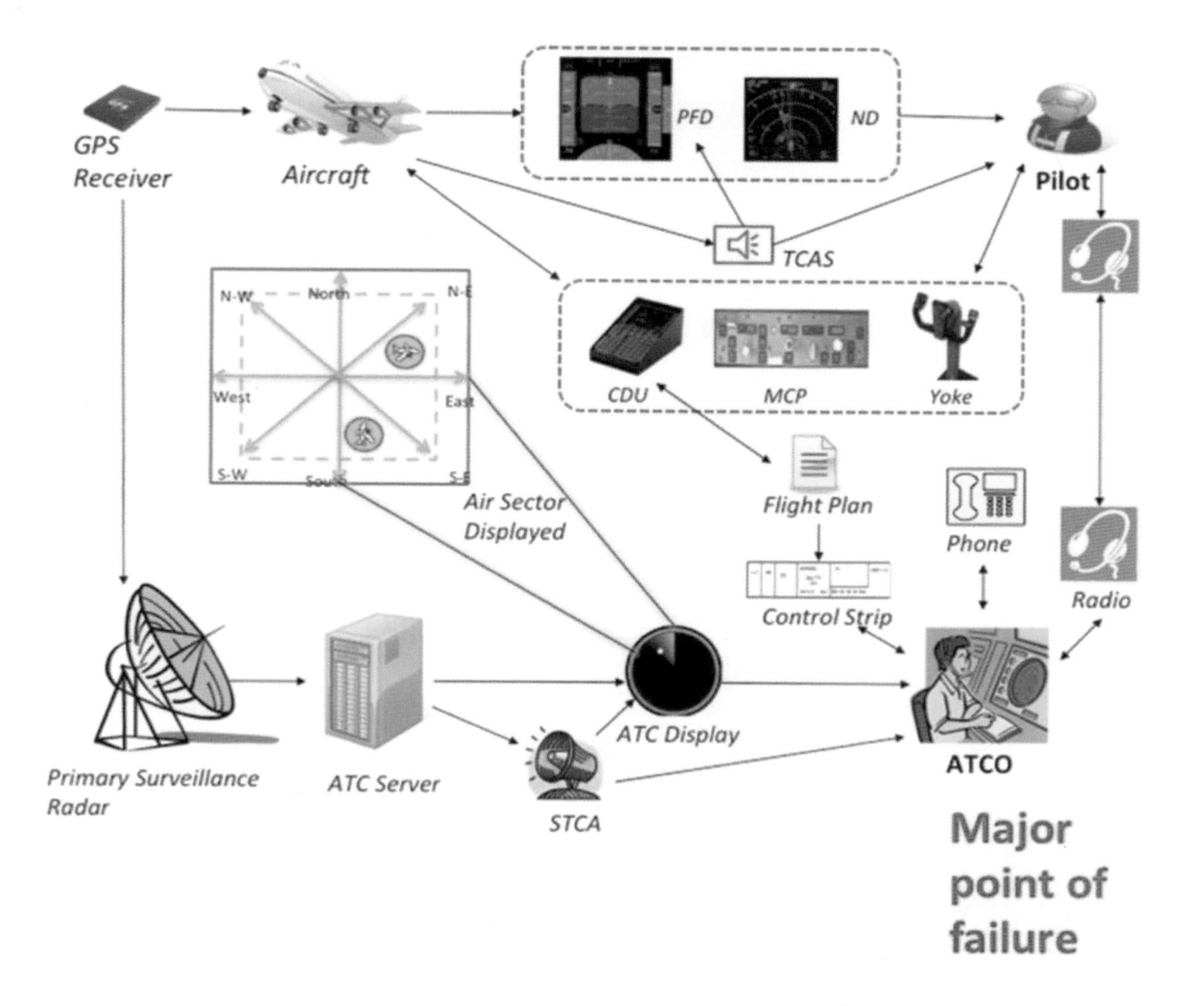


**Figure 16 ~ System Nodes and Communication Links of the Überlingen Air Collision**

Figure 16 looks like a complex system of systems but, using LEDSM as a safe communication development tool, it breaks down simply into four Enterprises:

- Zurich Air Traffic Control, (ATC)
- Aircraft that visit Friedrichshafen, (where an aircraft approaching becomes an Enterprise during landing, and again at take-off, but then disappears until another aircraft is taking off or landing when it too becomes an Enterprise; those are the only times during the night that Zurich ATC is required to support aircraft movements at Friedrichshafen)
- Bashkirian Airlines Flight 2937, a Tupolev Tu-154 passenger aircraft
- DHL International Aviation ME Flight 611, a Boeing 757 cargo aircraft

To establish the links between the Enterprises and within each Enterprise and the protocols involved, WBTR can be used. The WBTR will also establish what DGRs must be implemented to guarantee the best possible response and thus the most likely reduction risks. This, in a WBTR exercise would be used as the classic example of the envelope within which Zurich ATC must operate. However, in this exercise, because it is hypothetically done *a priori*, the fact that the two aircraft types and owners are now known from the actual accident, would not be relevant, they would simply be labelled 'Aircraft 1' and 'Aircraft 2'.

### *10.2.2 Using WBTR in a Hypothetical Pre-collision Review: Stage 1*

Here is an exemplar of the sort of reasoning that would help establish whether the communication system at Zurich is adequate or not and what needed to be done to improve the odds of avoiding a collision:

**Why** would two aircraft collide?

**Because** the avoidance procedures used were wrong. **Therefore**, all errors in procedures that could lead to a collision should be eliminated.

**Why** could current procedures be inadequate? (Note two replies to this but there may be more, these are just the relevant ones for the exercise).

**Because** they can be misinterpreted. **Therefore**, procedures must be unambiguous.

> **What** makes us prone to writing procedures that are inherently ambiguous?
>
> **Because** we have not fully explored the scope, we have assumed that procedures we know about are used universally. **Therefore**, the scope must be fully explored and the outputs from the examination of the extent of the scope incorporated in the procedures for universal adoption in air space under our control.
>
> > **Why** has the scope not been fully explored?
> >
> > **Because** there are always more issues emerging. **Therefore**, regular reviews of the scope of procedures and related behaviours must be undertaken.

**Because** at some periods the communication systems may be down for maintenance, or due to failure, and formal procedures to deal with those occasion have not yet been adopted. **Therefore**, maintenance procedures must include ensuring the workload can still be coped with and all personnel needing to know are informed.

> **Why** could a notified maintenance process lead to an unmanageable workload?
>
> **Because** procedures have not been formalised. **Therefore**, maintenance processes must be formalised to the extent that, as a minimum, signatures of involved parties when maintenance is to begin and when maintenance has ended must be collected as evidence of knowledge of any scheduled degradation of service.

### *10.2.3 Using WBTR in a Hypothetical Pre-collision Review: Stage 2*

The next stage is exploration of the procedures individually. For those managing the ATC, the following should have been apparent if time invested in LEDSM/WBTR review:

**Why** would current procedures for Air Traffic Controller behaviour be inadequate? (Four answers here)

1. **Because**, currently, rest periods are allowed for one of the two controllers but no method for the awake controller to raise an alert exists. **Therefore**, an alert should be installed.
2. **Because** the ATC is not formally made aware of when maintenance starts and ends. **Therefore**, a notification system should be put in place.
3. **Because** pilots may have procedures in direct opposition to another airline's procedures for TCAS, (Bashkirian prioritised ATC, DHL prioritised the TCAS). **Therefore**, global procedures must be put in place
4. **Because** a communication link to another ATC may be unusable. **Therefore**, at least one additional channel must be included.

It will usually be best practice to get two teams to examine a safety-critical communication system using WBTR, because, as can be seen, similar conclusions may come from their different approaches, but new insights may also be gained by slightly different questioning and reasoning.

#### *10.2.4 Using WBTR in a Hypothetical Pre-collision Review: Stage 3*

Other questions that may arise in discussions with a safety group looking at such a system might be:

**Why** would the ATC not hear the TCAS alert? (It was not heard on the night of the collision.)

**Why** would the land line telephone not work?

**Why** would the pilot disobey the TCAS instruction? **Because** some countries tell pilots to prioritise instruction from the ATC. **Therefore**, issue instruction to obey TCAS under all circumstances in governed airspace.

### 10.3 Using WBTR to Examine the Townsend Thoresen RoRo Ship Capsize

In hindsight, it would have been so easy to prevent the unnecessary deaths that occurred when the Townsend Thoresen ferry sank as it left Zeebrugge in 1987, had a WBTR approach been taken. WBTR analysis is efficient at identifying erroneous assumptions but, as has been said in the past, "*It does take a wit to ask a pertinent question*". Hence it is always advisable to employ safety experts in the particular field with proven abilities, SQEP, since those who will eventually use a system, while being skilful, may be over-confident that they would not make 'such a silly mistake'. The simple question, "**What** would cause the ferry to capsize?" appears not to have been asked or, if it was, not followed through in a logical manner to develop behaviour protocols for employees of the ferry company; protocols being an output of LEDSM.

All that was needed to be absolutely sure of the ferry never leaving port with the doors open was:

1. A direct link to the Boatswain to enable confirmation from him that the doors are closed.
2. An interlock detection system that provided a light on the Captain's desk showing the doors are fully closed.
3. A protocol sheet, much as airline pilots have check lists, that ensures the Captain will not set sail without all the required actions being completed.

## 11 The Benefits of LEDSM

### 11.1 Preamble

In summary, LEDSM will result in what have been commonly called Memoranda Of Understanding (MOUs). These will consist of affirmed Dependency/Guarantee Relationships. The protocol is a set of one or more guarantees that accurate data will be provided in a timely fashion to another that is dependent on that data. The associated techniques of system and design analysis ensure that the need for extra channels, where there is the potential for one communication channel to fail, is explored. LEDSM diagrams provide an easy to comprehend, visual expression of the connections at any one time.

Provided here is a list of some of the benefits of LEDSM used in conjunction with WBTR. LEDSM offers a number of benefits to those dealing with many complex safety-critical communication requiring systems. Primarily, potentially catastrophic situations can be averted by reducing miscommunication risks. The benefits are divided into three categories, End User, Investor/Management, and Safety Engineering.

- End Users of a data network will be able to gain situational awareness of the nature of the network that is producing their data and understand whether it is adequate. The ability to keep this awareness dynamically as the network morphs will be helpful.
- The Investors in, and Management of, the Enterprise will be able to understand the limitations and capabilities of the network to deliver essential safety-related data and information.
- Those developing systems with safety related aspects will find it effective in improving Safety Engineering and safety assurance of networked data driven systems, especially where there are humans in the loop.

### 11.2 End User Benefits of LEDSM

- Its graphic nature offers easy-to-understand confirmation to users that their needs are satisfied.
- Network graphics can be divided up, so that any individual need only focus on the part of the network relevant to their own role.
- It easily highlights omissions to those checking that they are in touch with who they need to receive or send data to.
- It is flexible, and any network can easily be expanded or reduced when new stakeholders come or leave (although, of course, it will usually be necessary to add in new expertise if an essential contributor to one's network leaves).
- Each level communicates directly to someone trained and capable in understanding the science and issues associated in other Enterprises, peer knowledge thus increases.
- It helps stakeholder Enterprise staff to identify their peer groups, and avoid delays in communication.
- Off-line (and for their own peace of mind), researchers, politicians, engineers, scientists, medics, and commercial businesses can generate their own maps of their networks at levels of granularity that are different from those handed down by protocols within their Enterprise. LEDSM thereby enables them to see how, with some thought applied, they can propose a case, for instance, to the higher levels vertically above them in their Enterprise, for further communication links to be formally established with Enterprises through horizontal protocols.

### 11.3 Investor/Management Benefits of LEDSM

- It is system neutral, being applicable in many situations.
- Network graphics from new relevant Enterprises are easily integrated.
- It facilitates the development of protocols that dictate the primary content of safe communication and in doing so reduces risks. These can be either vertical within an Enterprise or horizontal with other Enterprises. They facilitate the creation of rules that are important to prevent *laissez-faire* approaches to mapping communication needs for emergency situations.
- The LEDSM diagram can be condensed or expanded as required by the size of the Enterprise or the particular state of the project in its lifecycle.

- It provides a negotiating tool with which to demonstrate needs to recalcitrant provider stakeholders, and shows the benefits that they would receive.
- Enterprises needing data can identify missing information sent to them and will have already agreed on what basis they can proceed when data is less than the maximum desirable. Equally, data senders will be confident that the maximum information that can be extracted, from what they send, will be presented to all potential users without needing to be concerned that their data will be ignored as lacking detail, action levels having already been agreed in advance.
- LEDSM, WBTR and DGRs for these communication networks are low cost to implement.
- The use of LEDSM can reduce post disaster costs. Increasingly, it is the case that Post-Traumatic Stress Disorder (PTSD) leaves emergency workers and all those who have experienced traumatic events at close hand, with long term stress. The stress results in physical illness, mental health problems, many days of absence due to poor sleep and stress, and little option but to sue their employer for compensation. As has been shown, particularly in the simple case of the suicide bombing of the Ariana Grande concert, the probability of effective emergency responses is increased by planning using the LEDSM tool *a priori* to analyse possible situations and responses. Hence emergency personnel and victims will be less likely to experience ill-feelings of not having done enough for victims to get them necessary help. Thus, they will be less stressed after the event and thereby there will be less cost to the involved organisations in terms of compensation, court proceedings and lost days of work by affected employees.

## 11.4 Safety Engineering Benefits of LEDSM

- It overcomes the lack of, or poor, communication planning, which is so often responsible for accident deaths.
- It integrates the concepts of WBTR and DGRs to further reduce risks.
- It uses the principle of the proven technique HAZOPS to assist in identification of data flows through WBTR brainstorming.
- It facilitates the development between people, between people and systems, and between stand-alone systems, of DGRs. This secures a mutual understanding of needs and capabilities in safety-related situations.
- It helps to focus an Enterprise on their organisation's safety communication protocols to reduce unnecessary distraction, and to protect the integrity of proprietary data while being open on safety issues.
- When alerts are necessary, those first to identify the arising problem can rapidly disseminate information, having already identified in advance the peer group and the extent of data release permitted by their Enterprise.
- If a noted expert in a particular field passes on, moves on, or retires, the retention of a rapid alert system to deal with emerging threats is facilitated.
- It considerably lowers the level of risk associated with miscommunication and a lack of communication and therefore can be used as far as practicable, where "ALARP" (As Low As Reasonably Practicable) principles are expected.
- SQEP, i.e. Suitably Qualified and Experienced Personnel, will find LEDSM easy to understand and implement.
- The need for redundant communications and protocols to ensure extra channels are acquired, and used, is made obvious when risks for those supervising safety-critical operations are analysed. As a generalisation, all supervisors of safety-critical Enterprises should have more than one communication channel with supervisors of other Enterprises.

- It integrates stakeholder diagrams well with high-level system design, thus helping to eliminate omissions.

LEDSM used in conjunction with WBTR offers a way to cope with the enormous amounts of data, which will continue to increase, by offering a more-formal approach to development. Referring to the massive flow of COVID-19 information, misinformation, and disinformation influencing public health measures, it has been stated that, "*We are concurrently inundated with a global epidemic of misinformation, or an infodemic, primarily being spread through social media platforms; its effects on public health cannot be underestimated. Thus, the pandemic provides an opportunity to develop infodemic management approaches.*" (Sauer et al. 2021). LEDSM and WBTR maybe seen as one such infodemic management approach.

Time and time again we see that the problem in an accident or incident, where deaths and serious injuries have occurred, was the poor communication. Our systems have become safer and safer, and one only has to look at the huge reduction in airplane crashes over the decades to see that. This is the reason LEDSM and WBTR are proposed, to encourage owners of systems, especially those that involve human beings, to take a more-formal approach to developing the procedures that are to be put in place to prevent fatal accidents and emergency responses.

Let us all try to end this global epidemic of misinformation, disinformation and, especially for engineering safer emergency systems, the failure to provide accurate and timely information in emergency prevention and response systems involving humans.

## 12 Conclusions and Future Work

The Layered Enterprise Data Safety Model, LEDSM, offers a method of bringing the precision of electronic interface design to Emergency Service, Industrial Process, Engineering Design, and Military Command responses. Protocols also offer the precision of electronic safety-critical system development by mirroring how Dependency/Guarantee Relationships work. The use of WBTR, Why/What Because Therefore Reasoning, draws on hazard analysis and accident investigation techniques to establish design criteria in a formally documented way which may affect the design of a system of systems by reducing risks through the introduction of additional requirements. It is a technique that can easily be used iteratively when the relevant LEDSM or reports of usage indicate change is necessary.

Currently, an Enterprise would be expected to develop LEDSM diagram using standard graphic tools. However future work may include the development of a specific tool with drop down menus to assist the distribution of diagrammatic representations of expectations, roles and connections. That will, thus, enable rapid assimilation by personnel of the requirements of their roles, confidence in actions to be taken, and the protocols to follow; it will also facilitate adjustments of networks when change becomes necessary, or when tasks like maintenance need to have adaptations. Changes can be readily explained and graphically displayed to users in order to avoid poor outcomes from temporary disruption.

### Acknowledgments

The author thanks Mike Parsons and John Spriggs of the SCSC for their support and advice, and the anonymous reviewers for their thought provoking and helpful comments.

**References**

2002 Überlingen mid-air collision. (2022). In *Wikipedia*. https://en.wikipedia.org/wiki/2002_%C3%9Cberlingen_mid-air_collision. Accessed 20th June 2022.

Causalis. (2018). *Why-Because Analysis* Causalis Limited. https://www.causalis.com/20-analytics/10-WBA/. Accessed 20th June 2022.

Collins, D. (2021, July 25). *Manchester Arena terror attack: the lost two hours*. The Sunday Times. Available at: https://www.thetimes.co.uk/article/manchester-arena-terror-attack-the-lost-two-hours-0b923sd3p. Accessed 20th June 2022.

Dijkstra E. (1972). *The Humble Programmer*. The 1972 Turing Award Lecture, in Communications of the ACM 15 (10), October 1972: pp. 859–866

Dilmaghani, B. and Rao, R. (2009). *A systematic approach to improve communication for emergency response*. In Proceedings of the 42nd Hawaii International Conference on System Sciences, Waikoloa. IEEE. doi: 10.1109/HICSS.2009.39

Engineering Council. (2021). *Guidance on Sustainability for the Engineering Profession*. Available at: https://www.engc.org.uk/media/3555/sustainability-a5-leaflet-2021-web_pages.pdf. Accessed 20th June 2022.

Farrar J and Ahuja A. (2021). *Spike: The Virus vs. The People - the Inside Story*. Profile Books, London.

Faulkner A, and Nicholson M. (2020). *Data-Centric Safety: Challenges, Approaches, and Incident Investigation*. Elsevier.

Hales N. (2020). *Formalising Communication on Potentially Catastrophic Safety Projects*, In The Safety-Critical Systems Club Newsletter, Volume 28, Number 2. Available at: https://scsc.uk/scsc-158. Accessed 20th June 2022.

Hales N. (2021). *Data Safety in Virus Outbreaks: Lessons Learnt and Recommendations*, In Proceedings of the 29th Safety-critical Systems Symposium. Available at: https://scsc.uk/rp161.11:1. Accessed 20th June 2022.

Higginbotham A. (2019). *Midnight in Chernobyl*. Transworld Publishers.

HMS Sheffield (D80). (2022). In *Wikipedia*. https://en.wikipedia.org/wiki/HMS_Sheffield_(D80). Accessed 20th June 2022.

Huang J and Lien Y. (2012). *Challenges of emergency communication network for disaster response*. 2012 IEEE International Conference on Communication Systems (ICCS) pp. 528-532, doi: 10.1109/ICCS.2012.6406204

ISO/IEC. (1994). *Information technology — Open Systems Interconnection — Basic Reference Model: The Basic Model*. (ISO/IEC 7498-1:1994). ISO Geneva https://www.iso.org/

Kermit Tyler. (2021). In *Wikipedia.* https://en.wikipedia.org/wiki/Kermit_Tyler. Accessed 20th June 2022.

Manoj BS, and Baker AH. (2007). *Communication challenges in emergency response.* In Communications of the ACM, Volume 50, Issue 3 pp. 51-53, doi: 10.1145/1226736.1226765

Martin J. (2006). *The Meaning Of The 21st Century: A Vital Blueprint For Ensuring Our Future.* Riverhead Hardcover.

MS Herald of Free Enterprise. (2022). In *Wikipedia.* https://en.wikipedia.org/wiki/MS_Herald_of_Free_Enterprise. Accessed 20th June 2022.

Pervez F, Qadir J, Khalil M, Yaqoob T, Ashraf U, and Younis S. (2018). *Wireless technologies for emergency response: A comprehensive review and some guidelines.* IEEE Access, Volume 6 pp. 71814 - 71838. doi: 10.1109/ACCESS.2018.2878898

Sauer, M. A., Truelove, S., Gerste, A. K., and Limaye, R. J. (2021). *A Failure to Communicate? How Public Messaging Has Strained the COVID-19 Response in the United States.* Health security, 19(1), 65–74. https://doi.org/10.1089/hs.2020.0190

Schaper D. (2021, January 8). *Boeing To Pay $2.5 Billion Settlement Over Deadly 737 Max Crashes.* NPR: National Public Radio. Available at https://text.npr.org/954782512. Accessed 20th June 2022.

ScienceDirect. (2022). *Dependency Relationship.* Available at https://www.sciencedirect.com/topics/computer-science/dependency-relationship. Accessed 20th June 2022.

SCSC, Safety-Critical Systems Club. (2016). *Data Safety Guidance.* Version 1.3. Available at https://scsc.uk/scsc-127A. Accessed 20th June 2022.

SCSC, Safety-Critical Systems Club. (2019). *Data Safety Guidance* Version 3.1. The Abstract is available at https://scsc.uk/scsc-127D. Accessed 20th June 2022.

Shaw. S. (2022). *The OSI model explained and how to easily remember its 7 layers.* NetworkWorld. Available at: https://www.networkworld.com/article/3239677/the-osi-model-explained-and-how-to-easily-remember-its-7-layers.html. Accessed 20th June 2022.

Thomas M. (2005). *A View from the Stern.* Safety-Critical Systems Club Newsletter, Volume 14, Number 2, January 2005. Transcript available at https://scsc.uk/r77.2. Accessed 20th June 2022.

UKMoD. (2007). *Safety Management Requirements for Defence Systems.* (DEF.STAN.00-56/4). UK Defence Standardization, June 2007.

Why–because analysis: Example. (2006). In *Wikipedia.* https://en.wikipedia.org/wiki/Why%E2%80%93because_analysis#Example. Accessed 20th June 2022.

ISSN 2754-1118 (Online) — ISSN 2753-6599 (Print)

# Evaluating Software Execution as a Bernoulli Process

**Peter Bernard Ladkin**

Causalis Ingenieurgesellschaft mbH, Bielefeld, Germany

**Abstract**

*Daniels and Tudor consider evaluating software statistically through modelling execution as a Bernoulli Process (Daniels and Tudor 2022). They give examples and adduce considerations which they claim show that such modelling is "flawed". However, neither the examples they give, nor the considerations they adduce, follow the constraints necessary for modelling any process as a Bernoulli Process. I show that here.*

## 1 Introduction

### 1.1 Preamble

The statistical evaluation of safety-critical software is an important subject. I and colleagues have experienced over the last decade that this is one of the most misunderstood topics in safety-critical software evaluation. It also seems to be one of the most controversial; maybe these two phenomena are correlated. In a recent article, Daniels and Tudor claim that evaluating software execution using modelling as a Bernoulli Process is "flawed" (Daniels and Tudor 2022). However, neither their examples, nor the considerations they adduce, satisfy the constraints required for any Bernoulli Process modelling to succeed, as I show here. The appropriate conclusion would be, not that Bernoulli Process modelling is flawed, but that it is important to follow the constraints if you want trustworthy results.

The basic premiss of statistical evaluation is that one can observe a phenomenon occurring, note its characteristics, and analyse those data to see if there are any regularities. Those phenomena which do not exhibit obvious regularities can be said to behave stochastically. Even so, stochastic processes exhibit properties which can be characterised, such as the number of occurrences of a discrete phenomenon, or the relative frequency of values of some quantity. For many stochastic processes, there is an underlying probability distribution, as it is called, in which values of the characteristic can be taken to occur with a given probability.

Some probability distributions follow from very general, common characteristics of stochastic processes, and so occur ubiquitously; for example, the binomial distribution for discrete-valued events, and the normal (Gaussian) and Poisson distributions for continuously-valued events; there are others.

### 1.2 The Binomial Distribution in Mathematics and Engineering

The binomial distribution might underly a series of discrete events with exactly two discrete outcomes, which we can call success and failure. The binomial distribution occurs when there are repeated discrete events with one of two outcomes, in which the outcome of a later event is probabilistically independent of the outcomes of any earlier events, and in which the outcome of any event is given by a certain, fixed, probability. There are infinitely many binomial distributions, but when I specify:

- the probability $p$ of an individual event being a success; and
- the number of events $n$ I observe;

then I can say, through mathematical analysis, what proportion of the $n$ events I expect to result in success. I can also say what the chances are that some completely different proportion will be exhibited in my observations from that which I expect. Key to this is, of course, that each event indeed occurs with a given probability $p$. If this value varies over the events, then no dice; the events do not form a binomial distribution and I have to try something else.

Given that I have good qualitative reason to think that a certain kind of probability distribution is exhibited by a class of events, which is usually arrived at through prior consideration of how those events occur, then what I generally won't know beforehand are the values of the parameters. For a stochastic process, which I can argue qualitatively follows a binomial distribution, I can observe it for an appropriate number of events and attempt to estimate what I think the values of those parameters might be. The important parameter for the binomial distribution is the *fixed* probability of success. I need to be aware that what I observe may not be typical: stochastic processes are not deterministically law-governed, and even though I might expect a repeated coin toss (a prototypical example of a stochastic process taken to follow a binomial distribution) to yield me a roughly equal number of heads and tails on a long series of tosses, it just might give me all heads.

If all-heads is what I see, then it makes sense to ask:

- what is the possibility that I have an evenly-balanced coin, but that I just happened to get a run of all heads?
- what is the possibility that my coin is not at all evenly-balanced?

These are typically expressed as *confidence*[25]:

- Given the results, what is my confidence that the coin is evenly-balanced?
- Given the results, what is my confidence that the coin has a specific bias?

The even balance of a coin is represented by a probability $p$ of 0.5; the bias of a coin by some other value of $p$ between 0 and 1. Given my results, it turns out I can calculate my confidence in each of those circumstances (in each of the specific values of $p$).

This is all just maths, but here comes the engineering. Suppose a company is interested in coin-tossing. It manufactures coins for tossing. It claims to produce unbiased coins for tossing and it wishes to persuade its customers that it is very good at what it does. Then it can do a lot of tossing of each of the various coins it has produced, and, given the results, it can argue to customers:

- You can have confidence $X$ that a coin you bought from us is unbiased with an error of at most $\varepsilon$[26]

[25] "*Confidence*" is a technical term in statistical evaluation, usually expressed as a percentage from 0% to 100%, also known as "confidence level". This derives from the notions of "confidence interval" and "confidence set" (a confidence interval is a confidence set which consists of a continuous interval of numbers, for the case in which the value being estimated is a number). The notion of confidence set is explained in, e.g., (Siegrist, Chapter 7 Set Estimation, Section 1, Introduction); (Bedford and Cooke 2001, Section 4.3.3).

[26] "*Unbiased with an error of at most* $\varepsilon$" translates as "*p lies within the interval* $(\frac{1}{2} - \varepsilon, \frac{1}{2} + \varepsilon)$"

A customer can respond, "A*ll well and good, but your tests were all performed in summer, during business hours (0900-1700 UTC) and inside in windless conditions, and we are going to want to use your coin in winter during bar happy-hours of 1800-2000 outside in the wind in the beer garden. How should this affect the confidence?*" It is an important question, answered for example as follows:

- the time of day does not causally affect a coin toss in any plausible way;
- the season *per se* does not causally affect a coin toss, but the temperature in which the toss is performed modifies the coin (which is larger when it is warmer) as well as the atmosphere (which is less dense when it is warmer);
- the humidity of the atmosphere might affect the toss (humid air is less dense);
- wind may well affect a toss, as far as we can see.

Then there are possible causal effects which have not been introduced, for example the mechanical manner of the toss: if the translational velocity and rate of rotation are not stochastic but rather uniform, then such uniformity might causally affect the result.

You can assess these possible causal effects by performing the coin toss in the circumstances described and seeing what happens. The most the manufacturer can say is: "*Here are the circumstances (believed to be all the causal circumstances) in which we performed our tosses; if you perform your tosses in these circumstances, you will get the results with the confidence we said*".

We call these circumstances the *environment*, alternatively the *operational environment* or *operational profile*, of a stochastic process[27].

## 1.3 An Example from Safety Engineering

Here is an actual example of this reasoning in safety engineering. UK civil-nuclear regulations require that emergency systems are statistically evaluated. A SCRAM device is such a system; it causes an emergency shutdown of a nuclear reactor. A SCRAM device consists of various pieces of software-controlled hardware, plus communications, also to a large extent software-controlled nowadays. To a very high (specified) level of confidence, the SCRAM device must exhibit a very, very low (specified) failure rate (very, very low estimated value of $p$), in a given environment (specified). Through decades of experience and analysis, regulators, manufacturers, and operators (let us call these the operational stakeholders) believe they understand the environment very well.

Not only that, but the operational stakeholders believe they understand the characteristics of the hardware, and the characteristics of the communication hardware + software being used, so they can include these characteristics, if they wish, in the operational profile, and thereby apply the above estimation to the software alone.

Let us consider a theoretical SCRAM device. Actual SCRAM devices are similar to this, but somewhat more complicated in ways which for our purpose are uninteresting (that is, not causal to what we care about).

Sole input to the SCRAM device is a line which represents an "abnormal state" in some analogue fashion; if it remains "high" for a specific period of time, a SCRAM Command — a single high bit — is to be issued. The device is thus an arbiter. It has internal memory. After a SCRAM Command is issued, the device is completely reinitialised before being activated again: all registers and memory are reset to zero.

---

[27] Why all these different terms? Historically, different constituencies interested in statistical evaluation have developed their own terms, and it happens that these three are common in engineering.

The device may fail: all arbiters may fail (Lamport 2012).

Can we model this circumstance with a binomial distribution, as above? I think so (as does the nuclear industry in the UK, and many professional statisticians). Given the operational profile, starting in the initial state (all registers and memory and anything internally computationally causal are set to the specified initial state). The device is run until it provides an output, and the output recorded.

## 1.4 Trials, Bernoulli Trials and Bernoulli Processes

What does it mean that "*the device is run*"? Recall the two outputs: "success" and "failure". Does this describe what the device does? Not completely. There it is working in its operational environment. Suppose the line does not remain "*high for a specific period of time*". Then we would not want output to be a single high bit raised on the line. So there has to be some analysis of the device to develop confidence that this does not happen (to the given level of confidence). We don't follow this issue further here, because it is not best modelled as a discrete process — it is a continuous process-in-time with discrete occurrences of failure (when the output bit goes high without having experienced the trigger input). We are interested in the phenomenon that, if there is indeed an input trigger ("*line remains high for a specific period of time*") an output (high bit) is generated (*success*) or not (*failure*). That is a discrete process, with three components:

- A given starting state;
- The occurrence of a trigger input; and
- The output (success or failure).

When we have these three, this process forms what we may call a *trial*. A trial is an event with a beginning and an end, and an outcome. A particular class of interest here is that of *Bernoulli trials*. A succession of Bernoulli trials is called a Bernoulli Process (Siegrist, Chapter 10, Bernoulli Trials) (Ladkin 2017, Chapter 1, Software, the Urn Model, and Failure) (Ladkin and Littlewood 2016). Not every trial is a Bernoulli trial, and not every repeated trial forms a Bernoulli Process.

A Bernoulli Process is nothing more, nor less, than a sequence of trials of a process governed by the binomial distribution. It follows that there is reason to attempt to model a situation as a Bernoulli Process when there is reason to think that the underlying probability distribution is or might be binomial. Accordingly, the chances of success (failure) in a trial in a Bernoulli process must be the same for any trial (that $p$ in the binomial distribution); in particular, it must be independent of the history of the trials in the process. If you don't have reason to think that the underlying probability distribution can be represented as binomial, then you equally don't have reason to attempt to estimate the parameters of that distribution through treating the process as a Bernoulli Process.

## 1.5 A Trial Which is Not a Bernoulli Trial

Suppose the registers, memory, etc., are not reinitialised before performing the trial. The software may have some internal one-byte counter that is typically incremented three or four times in the course of a computation. If that counter starts at zero, the computation proceeds as one would expect. But suppose the counter has a value of 255. Then incrementing it will cause it to overflow and (let us suppose) trigger an exception condition. So, if the counter starts with a value of 255, then the chances of failure are 100% ($p = 0$). The probability of success of a trial with counter starting at zero is different from the probability of success of a trial with counter starting at 255. This situation does not fit a binomial distribution; such trials are not Bernoulli, and a sequence of them is not a Bernoulli Process.

## 2 Some Claimed Difficulties with Interpreting Software Execution as Bernoulli Trials

### 2.1 Preamble

If you are trying to model a situation as a Bernoulli Process, you first need to define what you take to be a trial, the events whose underlying distribution you are attempting to estimate; and, second, you need to argue that these trials are Bernoulli, that the underlying distribution is binomial. The probability concerned is the probability of that Bernoulli trial being successful.

Consider the way in which Daniels and Tudor introduce their topic. They speak of the "*probability of Success* [being] *the same every time the software is executed*" (Daniels and Tudor 2022, Sub-section 3.3.1). They don't define a trial; in that they don't define a starting state. The end state they consider is presumably the output "*Success*"; they don't address any concrete termination with an output of "*Failure*" — this leaves the possibility that software they are considering evaluating might not terminate. If that is the case, then there will be no trial. As they note, and as we have seen above, the interpretation of the probability (as imprecisely described above) taken at a snapshot in time can be very different, depending on the time and the internal state of the software at that time.

I see no plausibility in a suggestion that, taking executing software at a random point in its execution, and asking whether the computation will result in success, you can model this situation as a binomial distribution. What is "flawed" here is such a modelling attempt.

If you want software execution to be modelled as a Bernoulli Process, then you consider it as executing a series of Bernoulli trials, and it is the probability of success of the Bernoulli trial which is constant. If you identify a trial, but can't successfully argue that repeated trials of this sort are Bernoulli, then you are out of luck. Ladkin (2017, Chapter 1) shows that, very often, you can; but you do have to pay attention to the constraints of interpretation.

### 2.2 A Wrap-around Example

The example considered in Sub-section 1.5 above is isomorphic to the example considered in Daniels & Tudor (2022, Sub-section 3.3.2), with its counter wrap-around (in their example from 65535 to 0). Daniels and Tudor assert correctly that such a situation is not modelled as a Bernoulli Process in this way (Daniels and Tudor 2022). However, they do not define the situation they are modelling: they do not specify what is to count as a trial, let alone attempt to determine if it is Bernoulli.

The stipulated fixed probability $p$ in the binomial distribution has a mathematical consequence that Daniels and Tudor ignore in their discussion. The result of a trial of the system in Sub-section 1.5 is stochastically dependent on the previous trials: previous trials raised the value of the counter to 255 and resulted in $p = 0$ for this trial. In a Bernoulli Process, all trials are stochastically independent of one another; if they aren't, the process is not a Bernoulli Process.

That there is no Bernoulli Process at work here is not a subtle observation. This situation is well known to safety-critical system software developers — indeed, to most software developers. When you program a discrete function in software and you want to control its behaviour, you start it in a known starting state, usually by reinitialising (setting to zero) all memory and registers. Why? Because then the course of its computation will not be dependent on values created in previous computations. That is a good thing for software discrete-function programming. It is also a condition which must be assured if you want to evaluate a succession of calculations as a Bernoulli Process.

### 2.3 Summary of Conditions for Regarding Trials as Bernoulli

The discussion so far has illuminated the misleading introduction Daniels and Tudor (2022) give in their Sub-section 3.3.1 to attempting to model software execution as a Bernoulli Process.

- First, you must define what you take to be a trial with two possible but definitive outcomes.
- Second, it is not the "*probability of Success every time the software is executed*" that is relevant but rather the probability of success of the trial.
- Third, a trial must be stochastically independent of previous trials. If you don't have that, then the trials are not Bernoulli trials, and modelling as a Bernoulli Process cannot work.

The key is whether the underlying process plausibly follows a binomial distribution. If it doesn't, then attempting to estimate its parameters as if it were is unlikely to be fruitful.

### 2.4 Stepping Back and Looking

It is worth stepping back from the specific example of software execution for a moment to look at general characteristics of statistical evaluation, because some of the phenomena being highlighted have nothing to do with software *per se*.

Example: Suppose you are a Caucasian male of 60 years of age living in Western Europe, with no known impediments to your health. You are standing up, about to take two strides forward. What are your chances of dying while doing so?

- If you are standing on the unprotected parapet of a 50-story building, your chances are likely 100%.
- If you are standing at a tram stop, facing the tracks, your chances are arguably the period of time during which a tram is approaching the station and unable to stop before the point at which you cross its track, divided by the rest-of-the-time.
- If you are standing in your garden with only your lawn before you, the chances are arguably those of a person of your age and medical condition suffering an anomalous medical event (heart failure, aneurysm, etc.) within that period of time.

The first point to note is that the original situation is very much underspecified if you want to draw informative conclusions. The second point is that, in certain more specified circumstances, a probability can reasonably be assigned; indeed, it can reasonably be assigned in all these three cases.

So it is with the statistical evaluation of software; if the software is in an unknown state at the beginning of your trial, then the probability of its resulting in Success is equally all over the place. If the circumstances of its operation are sufficiently constrained according to known modelling requirements, then regularities can likely be identified and estimated.

This is well known to operators of computers. Back in the 1970's and 1980's, reboots of computer systems would succeed or fail, with outcomes not known deterministically to their operators, but known by "rule of thumb". Every tenth reboot would hang. Or every sixth re-boot. Every operator "knew" the proportion. I was involved in such proceedings. When the proportion changed, people would suspect something else was going wrong, and start looking at possible hardware weaknesses and so on. None (or, very few) of these operators were trained statisticians. They generally didn't know what a Bernoulli Process is (although the situation they were evaluating was arguably so modelled). They were just people with an experienced eye for regularities. That is what statistical analysis is — having an eye for regularities. If you can't see a binomial distribution before you, don't try modelling the situation as a Bernoulli Process. If you can and your data fits the Bernoulli Process assumptions, by all means try it and see.

On one point, Daniels and Tudor are very much correct. If you are running software which produces "*Success*", "*Failure*" output, you can't just pull out Table D.1 of IEC 61508-7:2010 Annex D, look up numbers and suppose without further consideration that they apply to your software[28]. More work is required to ensure that you are modelling your software execution appropriately.

## 2.5 Stepping Outside the Operational Profile

In their Sub-section 3.3.3, Daniels and Tudor consider the execution of software which fails (necessarily) on a leap-year date. They say "*Assuming there were no other defects, had we started running the program on 1 March 2016, then by 28 February 2020 we would have observed 35,040 hours of failure free operation. We would have had a high degree of confidence that the software is correct, yet it would have failed the very next day, 29 February 2020.*"

I have no idea how Daniels and Tudor might have obtained such a "*very high degree of confidence*", and neither should any other reader. They have not attempted to explain what model they are using which renders it. I also doubt they could plausibly have arrived at any such high degree. Construing an execution as a Bernoulli Process, or Poisson Process, or any other statistical evaluative model does not usually enable you to conclude with any degree of confidence, let alone a "*high degree of confidence*", what the outcome will be when the software executes with a hitherto unseen input.

Since (by hypothesis) the software in their example crashes on a leap-year date, it follows that the date is a causal input to the software. Let us consider such software. It is obvious that all dates on which the software has been observed to run are less than or equal to today, $T$. Assuming you don't know anything else about it, running software to which date is an input on a sequence of dates $<T$ enables no useful conclusion to be drawn on how the software will behave when date inputs are $>T$. Any "*high degree of confidence*" is misplaced.

It is a commonplace remark that statistical evaluations and estimates of future reliability can only occur for the operational profile which has been observed during statistical evaluation. When date is an input, the future operational profile is *disjoint* from the observed profile. It is hard to imagine deriving any useful conclusion with much confidence if the behaviour on the observed profile is all you know about the software[29].

---

[28] Anecdotes abound amongst safety-critical system engineering consultants of clients wanting to do this.

[29] In such a situation, it is usual to attempt to factor the software operation architecturally into a part dependent on date, and a part which is not, and the effects dependent on date be treated static-analytically. Such factoring may or may not succeed. Any reasonable statistical evaluation, however, depends on it succeeding at least in part.

I also note that such an example is not appropriately modelled using a binomial distribution. Daniels and Tudor are considering numbers of hours of failure-free operation. There is no notion of elapsed time which can be brought into rendering the software execution as a Bernoulli Process. The closest one can come is a parameter which is the number of Bernoulli trials. But number of trials is not necessarily related to elapsed time; in, say, 3 hours of operation, there might occur 0, 1, or 500 Bernoulli trials. If Daniels and Tudor want to reason about hours of failure-free operation, they cannot plausibly use Bernoulli Processes. A more appropriate model would be a Poisson Process or some other renewal process.

### 2.6 The Problem of Unknown/Inadvertent Input

As in the leap-year case above, it may indeed be that a particular software has inadvertent inputs that have not been recognised by the designers or users. If that is so, then when statistical evaluation if performed on an operational profile, these unrecognised inputs will likely be omitted from the explicit observed operational profile. Reasoning about the demonstrated reliability of the software will thereby be misleading.

This is a practical problem with statistical evaluation that must be addressed during any evaluation. Often, such inadvertent dependencies can arise through the use of library functions. It is said that much software is dependent on the date/time as well as on the availability of the Global Positioning System (GPS), which has no necessity to be dependent on those parameters, simply through using library functions which include them. For example, Hobbs recounts an incident in which GPS signals were jammed in the North Sea and a ship taken through to see what would happen (Hobbs 2015, p19[30]). Amongst other things, the ship's radar stopped working. "*The experimental team contacted the company that manufactured the radar and was told that GPS was not used within it. They then contacted the manufacturers of components in the radar and found that one small component was using GPS timed signals.*"

It is crucial that safety-critical system developers understand exactly what is input to their system. Building software which is dependent on unanalysed third-party functions from a library is generally unacceptable. Unanalysed dependency not only breaks accurate statistical evaluation, but much else besides.

### 2.7 Defect Clustering

The structure of this section parallels the sequence of examples in Daniels and Tudor Sub-section 3.3. In their Sub-section 3.3.4, Daniels and Tudor address software defect clustering; however, I don't see what that phenomenon might have to do with modelling software operation as a Bernoulli Process.

### 2.8 Non-operational Modes and Easter Eggs

In their Sub-section 3.3.5 Daniels and Tudor concern themselves with non-operational modes and Easter Eggs. Such modes may be undesirable for many reasons, but I don't see what problem they pose for statistical evaluation. Such modes have trigger input sequences. If the natural occurrence of such trigger input sequences is stochastic, then such executions indeed involve a fail of the functional intent of the software, but are

[30] Credit is given to Martyn Thomas.

thereby incorporated into any of the common statistical models (Bernoulli, or renewal). If the natural occurrence of the trigger sequences is not stochastic, then that indeed makes for difficulties in modelling.

### 2.9 Duration

In Sub-section 3.3.6, Daniels and Tudor address a phenomenon they call "*duration*". What they say, in full, is:

> *From a statistical point of view, it does not matter whether we run one copy of the software for a million hours or a million copies of the software for an hour each; both experiments result in one million hours of operation. In the real world, these two experiments are not the same. Running a million copies of the software for an hour each does not give us confidence that the software will run for two hours, let alone for days, months or years. Why do we expect the environment's behaviour for one hour to be the same as the environment's continuous behaviour for a million hours?*

If one is talking about software which executes over a period of time which is being measured, then a Bernoulli Process model is not likely to be appropriate, but rather a continuous-time model, for the reasons mentioned in Sub-section 2.5 above. But, if one nevertheless insists on trying to do so, then the closest pertinent parameter would be the number of executions of Bernoulli trials.

Translating what Daniels and Tudor say to the language of Bernoulli Processes, they would be contrasting running one copy of the software for a million trials with running a million copies of the same software for one trial each.

Both these options seem to me to define the same, or very nearly the same, Bernoulli Process. Daniels and Tudor want to claim that "*in the real world*" these two circumstances are different. They provide no argument for this contention. I don't agree with it. I think these two trial situations are the same in the pertinent respects.

## 3 The Upshot

On the basis of the critique exhibited in their Sub-section 3.3, Daniels and Tudor claim, "*The use of a Bernoulli process to model software failures is fundamentally flawed*" (Sub-section 3.6) and, "*We have shown in this paper that software reliability models that assume that software execution is a Bernoulli process are flawed*" (Sub-section 5).

I disagree that they have "*shown*" this. They have introduced some examples of faulty modelling and faulty statistical inference, which I have considered above in detail. The conclusion to draw is surely not that the statistical approach to evaluating software is flawed, but rather that if you apply a technique in various flawed ways then you are unlikely to get worthwhile results.

## 4 Reasoning With and About Probabilities

In their Sub-section 3.3.1, Daniels and Tudor speak about, "*The probability of Success* [being] *the same every time the software is executed*". This suggests that they think about probability as some objective characteristic inherent in the situation, in this case attaching to executing software. In the coin toss, similarly, one might think of the inherent probability of the unbiased coin turning up heads, as ½. This could be called the Laplacian conception of probability, as a property of an object exhibited in a circumstance.

Nowadays, no one takes probabilities to be such entities as this. Statisticians speak of the probabilities of events, of something happening (Newby 2020). Inductive logicians speak of the probabilities of sentences or assertions (Hacking 2001) (Howson and Urbach 2006) (Adams 1998). The Boolean logic of sets and propositions is more or less the same (complements, unions and intersections correspond to negations, disjunctions and conjunctions of propositions describing those sets), so one can correlate event classes with propositions, and vice versa. Probabilities in general are nowadays taken to be any assignment of (say) numbers between 0 and 1 to a Boolean algebra of sets or propositions which satisfy the Kolmogorov axioms (Hacking 2001). The idea of probability as being something inherent in a situation has lost much of the flavour it might have had to Laplace and the other pioneers. Probability nowadays is more like a collection of numbers which may be imposed on a situation according to certain constraints. An interpretation, not an inherence. There can be different interpretations for the same situations.

In fact, there are many contemporary conceptions of what probability might be. The two most common are the *frequentist*, in which a probability is a proportion which arises when an event is repeated an unbounded number of times, and the *Bayesian*, in which it represents something like a person's best guess, given original circumstances modified by experience with experiments (often called *subjective probability*).

In his introductory book on statistical evaluation, Spiegelhalter lists classical probability, 'enumerative' probability, 'long-run frequency' probability, propensity or 'chance', and subjective or "personal" probability as five different conceptions (Spiegelhalter 2019, pp 217-8). Furthermore, he says, *"...I take the view that any numerical probability is essentially constructed according to what we know about the situation — indeed probability doesn't really 'exist' at all (except possibly at the subatomic level)"*. Such a conception, that probability doesn't really exist at all, is *prima facie* hard to reconcile with such a concrete-seeming quantity as the "*probability of Success* [of] *software* [being] *executed*".

Spiegelhalter also has something to say about the phenomenon of misunderstanding in probability and statistics (Spiegelhalter 2019, p209):

> *"I am often asked why people tend to find probability a difficult and unintuitive idea, and I reply that, after forty years researching and teaching in this area, I have concluded that it is because probability really is a difficult and unintuitive idea. ... Even after my decades as a statistician, when asked a basic school question using probability, I have to go away, sit in silence with a pen and paper, try it a few different ways, and finally announce what I hope is the correct answer".*

Yup — been there, done that.

## 5 An Issue Not Addressed

Engineers reading Daniels and Tudor (2022), and then my commentary, might well be tempted to think that, if statistical reasoning is as tricky as this, then why don't we just avoid it altogether? Then no one will be misled. One answer is that, if we are not prepared to gather reliable data on software-based system performance and analyse it, there are pressing industrial problems which cannot be solved.

Consider Michael's Problem[31]. Michael's company produces sensors for safety-critical applications, typically for measuring the basic parameters of chemical and other industrial processes. The sensors are software-based, and haven't failed for software-related reasons in decades. They were not developed according to IEC 61508:2010 because that standard didn't exist when they were designed and built. But IEC 61508:2010 says that you can't use a piece of software in a new safety-critical application unless it is developed according to its precepts. The sensible engineering approach is surely to use the company's sensors *as they are* in a new application because they have proved themselves in use. A much less sensible engineering approach, but the approach nominally required by the standard, is to reimplement the software according to the precepts of IEC 61508:2010 and then use the extant hardware with the reimplemented software. You would be thereby replacing something known for a long time to be reliable with something with low operational experience that might do as well, or, alternatively, it might have a bug or two which need discovery and mitigation.

Avoiding statistical evaluation, as Daniels and Tudor advocate, will not and, maybe, cannot solve Michael's Problem.

## 6 Conclusion

Statistical evaluation of systems in their operation and the software which partly controls that operation is a task which naturally arises in engineering, for example in Michael's Problem.

But it seems to be tricky for many engineers; there appear to be misunderstandings and pitfalls. Some of these lie in the nature of the subject, but some of them seem to lie in misconceptions of statistical techniques.

I believe that enumerating the historical ways in which engineers have misused/attempted to misuse statistical reasoning in high-reliability environments is a worthwhile informational and cautionary exercise. To this end, a clear discussion of the examples raised by Daniels and Tudor in their Sub-section 3.3 is worthwhile. I hope I have contributed this.

**Acknowledgments**

Many thanks to Bev Littlewood, Martin Newby and Lorenzo Strigini for helpful comments.

**References**

Adams, E. W. (1998). *A Primer of Probability Logic.* CSLI Publications, 1998.

Bedford, T. and Cooke, R. M. (2001). *Probabilistic risk analysis: Foundations and methods*. Cambridge, UK: Cambridge University Press 2001.

Daniels, D. and Tudor, N. (2022). *Software Reliability and the Misuse of Statistics*, Safety-Critical Systems eJournal 1(1), SCSC-174, Safety-Critical Systems Club, January 2022. Available from https://scsc.uk/r174.3:1. Accessed 14th July 2022.

Hacking, I. (2001). *An Introduction to Probability and Inductive Logic*, Cambridge University Press, 2001.

---

[31] Named after a colleague who has it.

Hobbs, C. (2015). *Embedded Software Development for Safety-Critical Systems*, CRC Press 2015.

Howson, C. and Urbach, P. (2006). *Scientific Reasoning: The Bayesian Approach*, Third Edition, Open Court Press, 2006.

Ladkin, P. B. (2017). *A Critical-System Assurance Manifesto*, RVS-BI, 2017. https://rvs-bi.de/publications/RVS-Bk-17-01.html. Accessed 20th July 2022.

Ladkin, P. B. and Littlewood, B. (2016). *Practical Statistical Evaluation of Critical Software*. Paper presented at the 24th Safety-Critical Systems Symposium, Brighton, UK. Available from https://scsc.uk/r131/8:1. Accessed 14th July 2022.

Lamport, L. (2012). *Buridan's Principle*, Foundations of Physics 42 (8), August 2012, 1056-1066. Available from https://lamport.azurewebsites.net/pubs/pubs.html#buridan. Accessed 14th July 2022.

Newby, M. (2020). *Private discussion on calculating probabilities of Covid-19 infection*, 2020.

Siegrist, K. (no date). *Random: Probability, Mathematical Statistics, Stochastic Processes*. Available at https://www.randomservices.org/random/. Accessed 23rd July 2022

Spiegelhalter, D. (2019). *The Art of Statistics: Learning from Data*, Pelican Books, Penguin Random House UK, 2019.

# About the Safety-Critical Systems eJournal

## Purpose and Scope

This is the Journal of the Safety-Critical Systems Club CIC (SCSC), ISSN 2754-1118 (Online), ISSN 2753-6599 (Print). Its mission is to publish high-quality, peer-reviewed articles on the subject of systems safety.

When we talk of systems, we mean not only the platforms, but also the people and their procedures that make up the whole. Systems Safety addresses those systems, their components, and the services they are used to provide. This is not a narrow view of system safety, our scope is wide and also includes safety-related topics such as resilience, security, public health and environmental impact.

## Background

When the Safety-Critical Systems Club (SCSC) was set up thirty years ago, its objectives were to raise awareness of safety issues and to facilitate safety technology transfer. To achieve these objectives, the club organised events, such as Seminars and an annual Symposium, and published a newsletter, Safety Systems, three times a year.

The Newsletter has, in addition to news, opinion, correspondence, book reviews, and the like, also carried articles discussing current and emerging practices and standards. The length of such articles is limited to about two and a half thousand words, which does not allow an in-depth treatment. It was therefore been decided to add a third string to our bow and supplement the events and newsletter with this journal containing longer papers. The journal will be published here, as the Safety-Critical Systems eJournal, and is to comprise two issues a year.

## Content Sources

Sources include the outputs of SCSC working groups; solicited technical articles and topic reviews; submitted articles on new analysis techniques, discussion of standards, and industrial practice; and guidelines and lessons learned. If you wish to contribute, please see the Information for Authors[32].

Types of paper include, but are not limited to:

**Technical Articles:** Written by practitioners and describing practical safety assurance techniques and their industrial applications.

**Integration Studies:** Written by practitioners reporting upon successful (or otherwise) synergies achieved in practice with other assurance domains, e.g. security, environment, and resilience.

**Position Papers:** Written by, or on behalf of, Regulators, Standardisation Organisations, or other official bodies, setting out their position on a topic, e.g. the interpretation of a particular standard or regulation.

**Review Articles:** Papers highlighting recent developments and trends in some aspect of safety-critical systems or of their use in a particular industrial sector.

**Historical Articles:** Papers describing the development of safety assurance in an industrial sector; how we got to where we are today.

[32] https://scsc.uk/journal/index.php/scsj/information/authors

**Perspectives:** The authors' personal opinions on a subject, e.g. whether to use statistical methods in particular scenarios.

**Reports:** The lessons learned from incidents or the outcomes of trials with a description of scenarios, or methods, and a discussion of the results obtained.

**Working Group Outputs:** Written by Safety-Critical Systems Club Working Groups to include discussions, underpinning theory, or guidelines.

## Copyright and Disclaimer

# Index of Authors

# Index of Titles

Printed in Great Britain
by Amazon

17654658R00100